Signals and Communication Technology

This series is devoted to fundamentals and applications of modern methods of signal processing and cutting-edge communication technologies. The main topics are information and signal theory, acoustical signal processing, image processing and multimedia systems, mobile and wireless communications, and computer and communication networks. Volumes in the series address researchers in academia and industrial R&D departments. The series is application-oriented. The level of presentation of each individual volume, however, depends on the subject and can range from practical to scientific.

Indexing: All books in “Signals and Communication Technology” are indexed by Scopus and zbMATH

For general information about this book series, comments or suggestions, please contact Mary James at mary.james@springer.com or Ramesh Nath Premnath at ramesh.premnath@springer.com.

Shiban K. Koul · Somia Sharma

Microwave Integrated Circuits Based on Spoof Surface Plasmon Polariton

Springer

Shiban K. Koul
Centre for Applied Research in Electronics
Indian Institute of Technology Delhi
New Delhi, Delhi, India

Somia Sharma
Centre for Applied Research in Electronics
Indian Institute of Technology Delhi
New Delhi, Delhi, India

ISSN 1860-4862 ISSN 1860-4870 (electronic)
Signals and Communication Technology
ISBN 978-981-96-9040-4 ISBN 978-981-96-9041-1 (eBook)
https://doi.org/10.1007/978-981-96-9041-1

This Springer imprint is published by the registered company Springer Nature Singapore Pte Ltd.
The registered company address is: 152 Beach Road, #21-01/04 Gateway East, Singapore 189721, Singapore

To our Families

Preface

Microstrip line is the most popular technology used to design planar and flexible microwave integrated circuits because it is easy to fabricate, lightweight and planar. This line offers narrow bandwidth, high cross-talk and higher transmission loss. Spoof surface plasmon polariton (SSPP) based transmission lines can be used to cater to the disadvantages of the microstrip lines. The SSPP technology helps us to miniaturize the circuit size because of its subwavelength resolution; unwanted filtering circuits can be reduced as this line works as a low-pass filter, and cross-talk is less as the fields are confined in this case. This book bridges the gap between conventional microstrip technology, and the futuristic technology based on spoof surface plasmon polariton. The book presents detailed study of different components such as filters and antennas using spoof surface plasmon polariton. The book also focuses on biomedical applications of SSPP technology and its advantages. Nowadays, textile and flexible antennas are gaining popularity because of their applications in military and daily life wearable gadgets. These important aspects are studied in this book using SSPP technology.

The book is divided into nine chapters. Chapter 1 gives brief introduction of Surface Plasmon Polariton (SPP) and its mimic Spoof Surface Plasmon Polariton (SSPP). The focus is oriented towards use of SSPP in realizing microwave integrated circuits such as antennas, filters, couplers, and power dividers. SSPP offers several advantages as compared to existing conventional technologies and these are discussed in this chapter. Application of SSPP in microwave domain are also presented. Characterization of SSPP structures using dispersion diagram and equivalent circuit model is explained. Chapter 2 presents study of different type of transitions designed to convert conventional transmission line to SSPP transmission line (i.e. Quasi TEM mode to SSPP mode). The transitions that show excellent transmission efficiency over a wide frequency range and a good transition occupation ratio are discussed in this chapter. Different transitions from conventional transmission lines such as Microstrip lines, Coplanar Waveguides and Slot lines to SSPP transmission lines are studied in this chapter. Advantages and disadvantages of grounded and ungrounded structures are briefly presented in this chapter.

Filter design using spoof surface plasmon polariton and their advantages are discussed in Chap. 3. The types of filters that are studied in this chapter, are bandpass and band-notched filters. Different techniques, which are used to design bandpass and band-notched filters using spoof surface plasmon polariton, are discussed in this chapter. This chapter presents in detail advantage of realizing a spoof surface plasmon polariton-based filter. It helps one to gain insight into the need to use SSPP for filter design.

Chapter 4 explains tunable filter design using spoof surface plasmon polariton and its advantages. The type of filters studied in this chapter are tunable bandpass filters and tunable band-notched filters. Different techniques, which are used to achieve tunability in bandpass and band-notched filter using spoof surface plasmon polariton are explained in the chapter. This chapter presents detailed study high lightening advantages of designing spoof surface plasmon polariton-based tunable filters. Chapter 5 discusses frequency scanning antenna, which offers outstanding properties such as high directivity, wide frequency scanning and simple feeding mechanism. Different types of frequency scanning antennas such as travelling wave antenna and standing wave antenna are elaborated in this chapter. The method to achieve high gain and compact spoof surface plasmon polariton-based leaky wave antenna is explained. The standing wave antenna used to design a compact scanning antenna, is discussed in detail in this chapter.

In Chap. 6, we discuss planar endfire antenna design using spoof surface plasmon polariton with an improvement in gain and sidelobe level. Different techniques, which are used to design SSPP-based end-fire antennas, are elaborated in this chapter. Some of the applications of SSPP-based end-fire antennas are also discussed in this chapter. This chapter introduces the concept of circularly polarized end-fire antenna using SSPP. Chapter 7 elaborates the design and development of a power divider and phase shifter using spoof surface plasmon polariton. It discusses in detail the design of different types of power dividers such as simple T-section and Wilkinson power divider using spoof surface plasmon polariton. Spoof surface plasmon polariton-based phase shifters with different bit configurations and reconfigurability principles is discussed in this chapter.

In Chap. 8, applications and the advantages of the using spoof surface plasmon polariton in the Biomedical domain are explained. Different applications of spoof surface plasmon polariton in the field of biomedical domain such as cancer detection, tumour detection, biomedical imaging and bio-Sensing applications are discussed in this chapter. In Chap. 9, the circuit design on a flexible substrate or textile material using spoof surface plasmon polariton is elaborated. Why textile or flexible circuits are becoming necessary in our day-to-day life is an important point which is also discussed in this chapter. The advantages of spoof surface plasmon polariton technology over other conventional technologies is discussed in this chapter. There are different substrates or materials as well as printing technologies to design these circuits which are highlighted in detail in this chapter.

New Delhi, India

Shiban K. Koul
Somia Sharma

Competing Interests The authors have no competing interests to declare that are relevant to the content of this manuscript.

Contents

About the Authors

Shiban K. Koul is an Honorary Professor at the Indian Institute of Technology, Delhi. His research interests include RF MEMS, high-frequency wireless communication, microwave engineering, microwave passive and active circuits, device modelling, millimetre and sub-millimetre wave IC design, body area networks, flexible and wearable electronics, medical applications of sub-terahertz waves, EMI/EMC suppression techniques and reconfigurable microwave circuits including miniaturized antennas.

He has successfully completed 38 major sponsored projects, 52 consultancy projects and 62 technology development projects. He has authored 640 research Papers, 24 state-of-the art Books, 6 Book Chapters and 2 E-books. He holds 26 patents, 6 copyrights and 1 Trademark. He has guided 31 Ph.D. theses and more than 120 master's theses.

He is a Life Fellow of IEEE and Fellow of INAE and IETE. He is the Chief Editor of *IETE Journal of Research.* He served as a Distinguished Microwave Lecturer of IEEE MTT-S for the period 2012–2014. Currently, he is a Distinguished Lecturer of IEEE EMC-S for the period 2024–2025. He is recipient of numerous awards including IEEE MTT Society Distinguished Educator Award (2014), Teaching Excellence Award (2012) from IIT Delhi, Lifetime Achievement Award-2023 from the IETE, India, and Lifetime Achievement Award-2024 from Broadcast Society of India.

Somia Sharma received her B.Tech. degree in Electronics and Communication Engineering from Gautam Budh Technical University in 2011. She received M.Tech. degree in Radio Frequency and Microwave Engineering from Netaji Subhas University of Technology (NSUT) East Campus, formerly known as Ambedkar Institute of Advanced Communication Technologies and Research (AIACTR), New Delhi, in 2013. She served as an Assistant Professor in Krishna Institute of Engineering and Technology, Ghaziabad, from 2013 to 2017. She joined the Centre for Applied Research in Electronics, IIT Delhi, in January 2018 and defended her thesis in August 2023. She served as Guest Faculty in Jawaharlal Nehru University, New Delhi. She is working as Postdoctoral Fellow at the National Taiwan University, Taiwan, since July 2024. Her research interests include Terahertz Antenna, on chip circuits, spoof surface plasmon polaritons, frequency scanning antenna, tunable circuits and textile antenna. She is a Member of IEEE-Antenna Propagation Society and Microwave Theory and Techniques Society. She worked as Teaching Assistant for the lab courses (RF and microwave circuits and microwave measurement lab) and other theory courses including RF Active Circuits. She had held the post of Treasurer for year (2020) and Chairman for year (2021) in IEEE MTT-S Student branch chapter at IIT Delhi.

Abbreviations

ADS	Advanced Design System
AMC	Artificial Magnetic Conductor
BPF	Bandpass Filter
BST	Barium Strontium Titanate
CP	Circularly Polarized
CPW	Coplanar Waveguide
CSDMRs	Combined-Stub Dual-Mode Resonators
CSRRs	Complementary Split Ring Resonators
CSRs	Cross-Shaped Resonators
CST	Computer Simulation Technology
dB	Decibel
DG-RWG	Double-Grating Rectangular Waveguide
DGS	Defected Ground Surface
DLSSPPs	Double-Layered Spoof Surface Plasmon Polaritons
DSPSL	Double-Sided Parallel-Strip Line
ECC	Envelope Correlation Coefficient
ELC	Electric Field-Coupled
EM	Electromagnetic
EMI	Electromagnetic Interference
ESPPs	Effective Surface Plasmon Polaritons
ESSPPs	Effective Spoof Surface Plasmon Polaritons
FBR	Front-to-Back Ratio
FBW	Fractional Bandwidth
FOM	Figure of Merit
FPC	Flexible Printed Circuit
FPD	Filtering Power Divider
FSA	Frequency Scanning Antenna
FSM	Flexible Microwave
FSTSA	Frequency Scanning Tapered Slot Antenna
GaAs	Gallium Arsenide
GCFUs	Greek-Cross Fractal Units

GGW	Groove Gap Waveguide
GHz	Giga Hertz
GSIRs	Grooved Stepped Impedance Resonators
GSS	Graphene Sandwich Structure
HMSIW	Half-Mode Substrate Integrated Waveguide
HP	Horizontal Polarization
IDCLLRs	Interdigital Capacitance Loaded Loop Resonators
ISs	Interdigital Structures
ITU	International Telecommunication Union
LSPs	Localized Surface Plasmons
LSSP	Localized Spoof Surface Plasmon
LWA	Leaky Wave Antenna
MIMO	Multiple-Input-Multiple-Output
MMIC	Monolithic Microwave Integrated Circuit
MS	Microstrip
MSRR	Modified Split Ring Resonator
NWB	Narrow Wide band
OCCC	Ovarian Clear Cell Carcinoma
OECT	Organic Electrochemical Transistors
OLRs	Open Loop Resonators
OSB	Open Stop Band
PCB	Printed Circuit Board
PCMS	Periodic Corrugated Metallic Strips
PDMS	Polydimethylsiloxane
PEC	Perfect Electric Conductor
PET	Polyethylene Terephthalate
POCT	Point-of-Care Testing
Q-SSPPs	Quasi-Spoof Surface Plasmon Polaritons
QTEM	Quasi-Transverse Electromagnetic
RCS	Radar Cross Section
RF	Radio Frequency
RFID	Radio Frequency Identification
RWG	Rectangular Waveguide
SIPW	Substrate Integrated Plasmonic Waveguide
SIR	Step Impedance Resonator
SIW	Surface Integrated Waveguide
SOC	Serous Ovarian Cancer
SPIs	Spoof Plasmonic Interconnects
SPP	Surface Plasmon Polariton
SRRs	Split Ring Resonators
SSL	Suspended Strip Line
SSPP	Spoof Surface Plasmon Polariton
S-SRRs	S-Shaped Split Ring Resonators
SUT	Sample Under Test
SWLWA	Surface Wave Leaky Wave Antenna

TARC	Total Active Reflection Coefficient
TEM	Transverse Electromagnetic
TFMF	Tunable Fabric Microwave Filter
TL	Transmission Line
TM	Transverse Magnetic
UHF	Ultra-High Frequency
UWB	Ultra-Wideband
VNA	Vector Network Analyser
VP	Vertical Polarization
VSRR	Vertical Split-Ring Resonator
VSWR	Voltage Standing Wave Ratio
WBANs	Wireless Body Area Networks
WGM	Whispering-Gallery Mode
Wi MAX	Worldwide Interoperability for Microwave Access
WLAN	Wireless Local Area Network

Chapter 1
Introduction to Spoof Surface Plasmon Polariton

1.1 Surface Plasmon Polariton

Surface plasmon polaritons (SPPs), which exist at optical frequencies at the interface of two mediums with opposite permittivity, are highly localized surface waves [1]. Although the SPP operates exceptionally well at optical frequencies, metal behaves as a perfect electric conductor (PEC) at microwave frequencies, making them impossible to produce naturally. In the optical regime, the incident wave interacts with the electron plasma close to the metal's surface, which in turn causes widespread oscillations to spread along the interface. Due to the metal's strong conductivity, the electric vector needs to have a component that is normal to the interface. Surface plasmons (SPs) are the highly localized modes at the surface of two materials (e.g., metal–dielectric interface) of opposite signs in the real parts of the electric permittivity. These waves propagate along the interface of a conductor and decay exponentially in the transverse direction [2]. SPs reduce the effective wavelength significantly and therefore present high spatial confinement [3]. Ebbesen et al. [4] reported extraordinary transmission in optical wavelengths, through subwavelength apertures owing to the coupling of light with surface plasmons (SPs).

SPs have been found to have potential uses in a variety of fields, including photovoltaics [5], nanophotonic [6], biomedical sensing (fiber sensors) [7], optoelectronics [8], and optical signal processing [9]. Figure 1.1 shows that SPP waves fundamentally behave like TM waves since they propagate along the *x*-axis. According to Maxwell's equation [2, 3], the magnetic and electric fields in upper space, i.e., for ($z > 0$) are

$$H^{+} = (0, A, 0)\mathrm{e}^{(-j\beta x - \alpha_1 z + j\omega t)} \tag{1.1}$$

$$E^{+} = (-jk_1, 0, \beta)\frac{A}{\omega\varepsilon_1}\mathrm{e}^{(-j\beta x - \alpha_1 z + j\omega t)}. \tag{1.2}$$

S. K. Koul and S. Sharma, *Microwave Integrated Circuits Based on Spoof Surface Plasmon Polariton*, Signals and Communication Technology,
https://doi.org/10.1007/978-981-96-9041-1_1

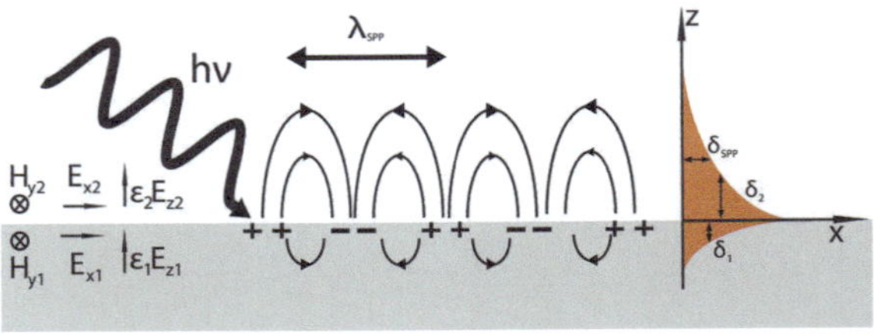

Fig. 1.1 Electromagnetic field on a metal–dielectric interface where surface plasmon polariton exists

In lower space, i.e., for ($z < 0$)

$$H^{-} = (0, B, 0)e^{(-j\beta x+\alpha_2 z+j\omega t)} \tag{1.3}$$

$$E^{-} = (-jk_2, 0, \beta)\frac{B}{\omega\varepsilon_2}e^{(-j\beta x+\alpha_2 z+j\omega t)}. \tag{1.4}$$

Here $\beta = k_x$; the wave number in the propagation direction and the decay factor α_1 and α_2 can be calculated by

$$\alpha_1 = \sqrt{\beta^2 - \varepsilon_1(\omega/c)^2} \tag{1.5}$$

$$\alpha_2 = \sqrt{\beta^2 - \varepsilon_2(\omega)(\omega/c)^2}. \tag{1.6}$$

Re (α_1) and Re (α_2) must be positive for the E-field and H-field to decrease exponentially and ensure substantial confinement of the electromagnetic field. Transverse components of the magnetic and electric fields are continuous at the contact when boundary conditions are applied. Surface plasmon condition is obtained as

$$\frac{\alpha_1}{\alpha_2} = -\frac{\varepsilon_2(\omega)}{\varepsilon_1}. \tag{1.7}$$

According to (1.7), for SSPPs to be localized to the interface, $\varepsilon_2(\omega)$ should be negative.

The simplified form of the Drude model of the metal is given as

$$\varepsilon_2(\omega) = 1 - \frac{(\omega_p^2)}{\omega^2}, \tag{1.8}$$

where $\varepsilon_2(\omega)$ is the frequency-dependent dielectric function of the metal, ω_p is the plasma frequency of the metal, and ω is the operating frequency of the electromagnetic wave. Metal exhibits a frequency-dependent dielectric function with $\varepsilon_2(\omega) = \varepsilon'(\omega) - j\varepsilon''(\omega)$ which follows the Drude–Lorentz model and hence possesses a negative real part below its intrinsic plasma frequency (which occurs in the ultraviolet region of the spectrum); hence, SPP occurs below the ultraviolet frequency. From (1.5)–(1.7), the dispersion relation obtained is given below:

$$\beta = k_0 \sqrt{\frac{\varepsilon_1 \varepsilon_2(\omega)}{\varepsilon_1 + \varepsilon_2(\omega)}}. \tag{1.9}$$

Two important aspects of SPP are

(1) Dispersion curve is always right to the light line which indicates SPP is a slow wave with a propagation constant larger than that of free space. At low frequencies, the propagation constant decreases rapidly, and surface plasmon frequency accounts for highly localized SPP.
(2) SPP propagates along the interface from zero frequency with the same group velocity as that of light and slows down as the frequency increases and asymptotically becomes localized when the frequency approaches the surface plasmon frequency.

1.2 Spoof Surface Plasmon Polariton

SPPs exist only in the visible and near-infrared frequencies in which metal and dielectric material have opposite permittivity below the plasma frequency. As the frequency decreases to microwave or terahertz frequency, the metal behaves as a perfect electric conductor rather than a plasma with negative permittivity and hence does not support SPP modes. Regardless of its excellent performance at optical frequencies, SPP cannot be generated naturally at microwave frequencies. Periodic subwavelength structures are used to mimic the surface plasmon at microwave and terahertz regimes known as spoof surface plasmon polariton. A spoof surface plasmon is a type of surface wave that exists at the metal–dielectric interface.

In 2004, Pendry et al. demonstrated the SSPP behavior. He considered an array of holes perforated on a perfect conductor with a square cross-section and took smaller spacing compared to the wavelength and analytically proved that the dispersion behavior of the structure was like that of SPP [10]. Pendry et al. have also analytically calculated the dispersion relation when the metallic surface is corrugated with 2D hole having a square cross-section, with dimensions 'a' and period 'd' where both 'a' and 'd' are much smaller than the free space wavelength, as given in [10]. The holes are filled with a material having a dielectric permittivity ε_h and permeability μ_h. The dispersion behavior of the modes near the corrugated metallic layer is like the real SPP and the relation is given as

$$K_{II}^2 = \frac{\omega^2}{C_0^2}\left(1 + \frac{1}{\omega_p^2 - \omega^2}\frac{64a^4\omega^2}{\pi^4 d^4}\right) \tag{1.10}$$

where ω_p is

$$\omega_p = \frac{\pi c_0}{a\sqrt{\varepsilon_h \mu_h}}.$$

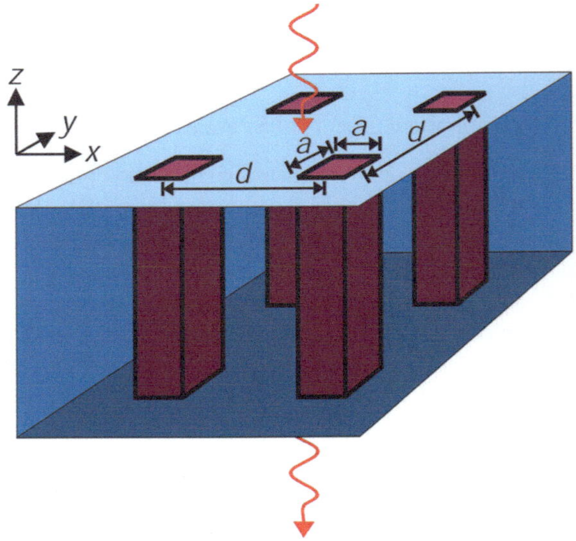

Fig. 1.2 Structured metallic surface [10]. Reproduced with permission from The American Association for the Advancement of Science

In 2005, Garcia-Vidal et al. [11] analytically proved that by inserting a corrugation with periodicity 'd', width 'a', and depth of corrugation 'h', we can achieve the property of surface plasmon at low frequency. The dispersion relation can be obtained by (Figs. 1.2 and 1.3)

$$\frac{\sqrt{(k_x^2 - k_0^2)}}{k_0} = \frac{a}{d}\tan(k_0 h). \tag{1.11}$$

According to Harington [12], a reactive boundary tends to produce wave guidance along that boundary. A reactive surface can be obtained by coating a conductor with a dielectric layer or by corrugating a conducting surface. We assume that the teeth are infinitely thin and there are many slots per unit wavelength. The teeth will short out any E_y permitting only E_z and E_x at the surface. TM field of dielectric slab is of this type; hence, we shall assume that this field exists in the air region (Fig. 1.4).

Maxwell's equations in homogeneous source-free region are given as follows:

$$\begin{aligned} \nabla \times E = -j\omega\mu H\hat{z} \quad \nabla.H = 0 \\ \nabla \times H = j\omega\varepsilon E\hat{y} \quad \nabla.E = 0. \end{aligned} \tag{1.12}$$

In view of the divergence less character of E and H, we can express the field in terms of magnetic vector potential (A) or in terms of an electric vector potential (F).

$$H = \nabla \times A \tag{1.13}$$

$$\nabla \times E = -j\omega\mu(\nabla \times A) \tag{1.14}$$

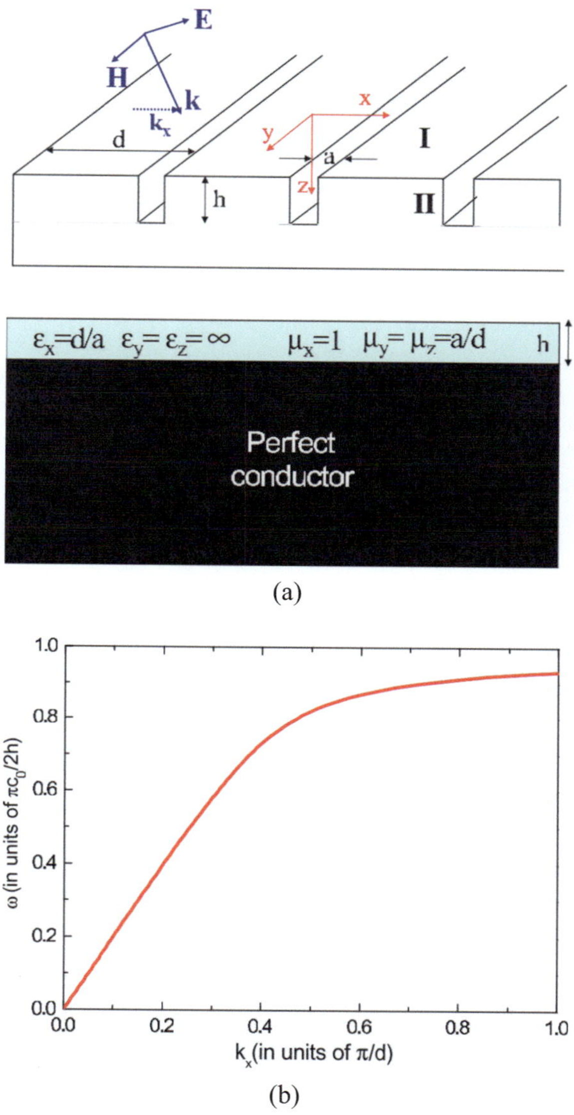

Fig. 1.3 **a** Structured metallic surface, **b** dispersion diagram for structured metallic surface calculated from (1.11) [11]. Reproduced with permission from IOP publishing

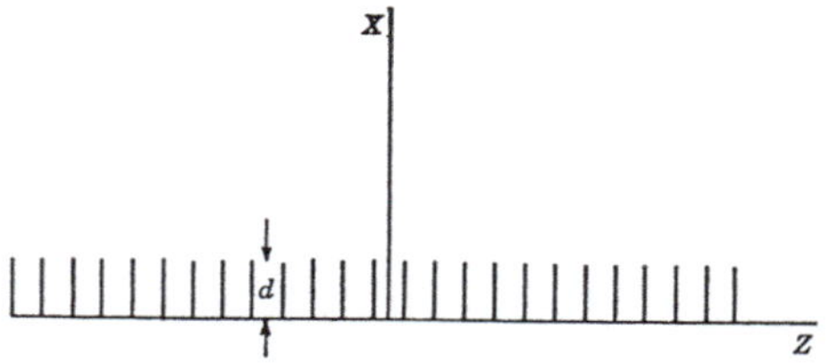

Fig. 1.4 Corrugated structure

$$\nabla \times (E + j\omega\mu A) = 0. \tag{1.15}$$

Any curl free vector is gradient of some scalar.

$$E + j\omega\mu A = -\nabla\varphi^a. \tag{1.16}$$

where φ is electric scalar potential

$$\nabla \times \nabla \times A - k^2 A = -j\omega\varepsilon\nabla\varphi^a \tag{1.17}$$

$$\nabla \times \nabla \times F - k^2 F = -j\omega\varepsilon\nabla\varphi^b \tag{1.18}$$

$$E = -\nabla \times F + \frac{1}{j\omega\varepsilon}(\nabla \times \nabla \times A) \tag{1.19}$$

$$H = -\nabla \times A + \frac{1}{j\omega\mu}(\nabla \times \nabla \times F) \tag{1.20}$$

$$\nabla . A = -j\omega\varepsilon\varphi^a \tag{1.21}$$

$$\nabla . F = -j\omega\mu\varphi^b. \tag{1.22}$$

For TM Wave, $F = 0$, and after solving the Maxwell's equation for $x > d$, the field equations are

$$E_x = \frac{k_x}{\omega\varepsilon_0} H_y \tag{1.23}$$

$$E_z = \frac{-B}{j\omega\varepsilon_0} v^2 e^{-vx} e^{-jk_x x} \tag{1.24}$$

$$H_y = Bve^{-vx} e^{-jk_x x} \tag{1.25}$$

where $-v^2 + k_x^2 = k_0^2$

$$Z_{-x} = -\frac{E_z}{H_y} = \frac{jv}{\omega\varepsilon_0}. \tag{1.26}$$

This is inductively reactive so the interface must be an inductively reactive surface to support such a field. It is assumed that the parallel-plate transmission line mode exists inside the corrugation slots. These are short-circuited transmission lines of characteristic wave impedance, η_0. Hence, the input wave impedance can be represented by

$$Z_{-x} = j\eta_0 \tan k_0 d \tag{1.27}$$

For $k_0 d < \pi/2$, it will be inductively reactive. After equating, we have

$$v = k_0 \tan(k_0 d) \tag{1.28}$$

and

$$k_x = k_0\sqrt{1 + \tan^2(k_0 d)}. \tag{1.29}$$

The average wave impedance will be calculated in case of finite width of the teeth.

$$k_x = k_0\sqrt{1 + \left(\frac{g}{g+t}\right)^2 \tan^2(k_0 d)}, \tag{1.30}$$

where g = gap, t = width of the teeth, and d = depth of the corrugation. Equation (1.30) resembles the dispersion relation of the surface plasmon polariton.

1.3 Why Is SSPP the Future of Microwave Integrated Circuits?

Since microstrip line is a widely used transmission line at microwave and millimeter-wave frequencies, it is easy to fabricate. It has drawbacks of narrow bandwidth, low gain when used as antenna, larger transmission line loss, and susceptibility to crosstalk. On the other hand, SSPP has excellent transmission characteristics and can replace microstrip lines. High-speed circuits and systems can be miniaturized using SSPP technology and unwanted filtering circuits can be removed [13]. In the SSPP transmission line, by varying the corrugation depth, the dispersion characteristics can be varied which exhibits a corresponding cut-off frequency with a sudden decline in the transmission spectrum beyond the cut-off frequency. This helps in filtering purposes and miniaturization of the device. Transmission line loss is the next important feature of any transmission line. It is a difficult task to control the line losses of a microstrip line. Whereas in SSPP, these losses can be controlled by altering the wave number. Zhang et al. [1] demonstrated that an SSPP transmission line has low loss compared to the conventional microstrip line over a wide frequency range. Wu et al. [14] demonstrated that the EM fields are tightly confined inside the corrugated transmission line compared to that in conventional microstrip lines. Therefore, crosstalk between the two transmission lines is less in SSPP lines [14]. These transmission lines have the property of slow wave structure. It has been shown that the SSPP waves were trapped at different locations at different frequencies in the case of ultra-thin corrugated metallic strips [15]. SSPP-based structures are being used by researchers to develop low-loss, low-cost, and reliable transmission lines, components, and systems to be operated at microwave and millimeter ranges.

1.4 Applications of SSPP in Microwave Domain

Various SSPP-based structures have been used to realize passive components such as filters, antennas, and power splitters. The leaky-wave antennas (LWAs) based on SSPP have gained lot of attention. The SSP TLs used in the realization of LWAs have gained attention due to their high field confinement, low loss, and low profile [16, 17]. The above properties are used to realize leaky-wave antennas for wideband scanning [18–28]. Several pattern reconfigurable antennas [29], millimeter-wave antennas, end-fire antennas, circularly polarized antennas, and Low RCS antennas [30–42] are designed using the SSPP. Presently, a lot of work is going on in the field of passive and active components using SSPP [43–64].

1.5 Characterization of SSPP Structures

The spoof surface plasmon polariton structures inherit the properties of surface plasmon polariton such as dispersion characteristics, field confinement, and subwavelength resolution. The first step to designing any structure using spoof surface plasmon polariton is to design the unit cell and study its characteristics such as asymptotic cut-off frequency using a dispersion diagram. The geometric parameters of the grooves have a significant impact on the properties of the SPP TLs, such as transmission and dispersion. The characteristic impedance can be evaluated by an equivalent circuit model.

1.5.1 Dispersion Diagram

The Eigenmode solver is used to design the dispersion diagram of the spoof surface plasmon polariton unit cell. The dispersion characteristics depend on the geometrical parameters of the unit cell such as periodicity, depth, and shape of the unit cell. The dispersion diagram helps to study different modes supported by the unit cell. Higher-order modes of the unit cell are determined by the dispersion diagram. The dispersion relation for spoof surface plasmon polariton is shown in (1.30).

1.5.2 Equivalent Circuit

The equivalent circuit of an SSPP unit cell can be determined using transmission line theory with the help of which we can determine the characteristic impedance and the transmission characteristics which are studied in detail later in this book.

1.6 Outline of the Book

The book contains a total of nine chapters with the first chapter serving as an introduction to spoof surface plasmon polariton in the microwave domain where an overview of surface plasmon polariton and spoof surface plasmon polariton is given followed by their applications and characteristics of spoof surface plasmon polariton in microwave domain. Chapter 2 starts with a discussion on different transitions used to convert quasi-TEM or TEM mode to SSPP mode. The transition studied in this chapter covers all conventional transmission lines such as microstrip lines, coplanar waveguides, slot lines, and rectangular waveguides. This chapter also discusses the transition for both grounded as well groundless structures. Chapter 3 discusses the design, development, and complete characterization of bandpass filters and band-notched filters. The motivation is to observe the enhanced characteristics with the use of spoof surface plasmon polariton. Chapter 4 discusses the design and implementation of tunable bandpass filters and tunable band-notched filters. This chapter discusses how with the use of spoof surface plasmon polariton the tunability of the filters can be improved. Chapter 5 presents the design and implementation of a frequency scanning antenna using spoof surface plasmon polariton. The outstanding properties of frequency scanning antennas such as high directivity, wide frequency scanning, and simple feeding structure are discussed in this chapter. Chapter 6 reports the theory, design, and analysis of end-fire antenna using spoof surface plasmon polaritons. It also discusses circularly polarized end-fire antenna and its applications. Chapter 7 describes the need for a beam steering antenna and its application. In this chapter, different beam steering techniques such as electronic and mechanical are discussed. Reconfigurable phase shifters using spoof surface plasmon polariton for beam steering antenna are described in this chapter. Chapter 8 discusses spoof surface plasmon polariton for biomedical applications. In this chapter, different biomedical applications are discussed such as tumor and cancer detection, biomedical sensing, and imaging. Chapter 9 presents the design of the textile and flexible circuits using the spoof surface plasmon polariton. This chapter also discusses the need and advantage of spoof surface plasmon polariton for textile or flexible circuits.

References

1. Tang, W.X., Zhang, H.C., Ma, H.F., Jiang, W.X., Cui, T.J.: Concept, theory, design, and applications of spoof surface plasmon polaritons at microwave frequencies. Adv. Opt. Mater. **7**(1), 1800421 (2019)
2. Maier, S.A.: Plasmonic: Fundamentals and Applications. Springer, Berlin (2010)
3. Raether, H.: Surface Plasmons on Smooth and Rough Surfaces and on Gratings. Springer, Berlin, Heidelberg (1988)
4. Ebbesen, T., Lezec, H., Ghaemi, H., et al.: Extraordinary optical transmission through subwavelength hole arrays. Nature **391**, 667–669 (1998)
5. Polman, A., Atwater, H.A.: Photonic design principles for ultrahigh-efficiency photovoltaics. Nat. Mater. **11**(3), 174–177 (2012)
6. Ozbay, E.: Plasmonics: merging photonics and electronics at nanoscale dimensions. Science **311**, 189–193 (2006). https://doi.org/10.1126/science.1114849
7. Tsai, W.H., Tsao, Y.C., Lin, H.Y., Sheu, B.C.: Cross-point analysis for a multimode fiber sensor based on surface plasmon resonance. Opt. Lett. **30**(17), 2209–2211 (2005)
8. Cai, W., Shin, W., Fan, S., Brongersma, M.L.: Elements for plasmonic nanocircuits with three-dimensional slot waveguides. Adv. Mater. **22**, 5120–5124 (2010)
9. Gramotnev, D., Bozhevolnyi, S.: Plasmonics beyond the diffraction limit. Nat. Photonics **4**, 83–91 (2010)
10. Pendry, J.B., Martin-Moreno, L., Garcia-Vidal, F.J.: Mimicking surface plasmons with structured surfaces. Science **305**(5685), 847–848 (2004)
11. Garcia-Vidal, F.J., Martin-Moreno, L., Pendry, J.B.: Surfaces with holes in them: new plasmonic metamaterials. J. Opt. A **7**(2), 97–101 (2005)
12. Harrington, R.F.: Time-Harmonic Electromagnetic Fields. McGraw-Hill (1961). ISBN 9780070267459
13. Zhang, H., Cui, T.J., Zhang, Q., Fan, Y., Fu, X.: Breaking the challenge of signal integrity using time domain spoof surface plasmon polaritrons. ACS Photonics **2**(9), 1333–1340 (2015)
14. Wu, J.J., Hou, D.J., Liu, K., Shen, L., Tsai, C.A., Wu, C.J., Tsai, D., Yang, T.-J.: Differential microstrip lines with reduced crosstalk and common mode effect based on spoof surface plasmon polaritron. Opt. Express **22**, 26777–26787(2014)
15. Yin, J.Y., Ren, J., Zhang, L., et al.: Microwave vortex-beam emitter based on spoof surface plasmon polaritron. Laser Photonics Rev. **12**, 1600316 (2018)
16. Zhang, Q., Zhang, Q., Chen, Y.: Spoof surface plasmon polariton leaky-wave antennas using periodically loaded patches above PEC and AMC ground planes. IEEE Antennas Wirel. Propag. Lett. **16**, 3014–3017 (2017)
17. Tang, X., Zhang, Q., Hu, S., Kandwal, A., Guo, T., Chen, Y.: Capacitor-loaded spoof surface plasmon for flexible dispersion control and high-selectivity filtering. IEEE Microw. Wirel. Compon. Lett. **27**(9), 806–808 (2017)
18. Wang, M., Ma, H.F., Zhang, H.C., Tang, W.X., Zhang, X.R., Cui, T.J.: Frequency-fixed beam-scanning leaky-wave antenna using electronically controllable corrugated microstrip line. IEEE Trans. Antennas Propag. **66**(9), 4449–4457 (2018)
19. Kong, G.S., Ma, H.F., Cai, B.G., Cui, T.J.: Continuous leakywave scanning using periodically modulated spoof plasmonic waveguide. Sci. Rep. **6**, 29600 (2016)
20. Hao, Z.-C., Zhang, J., Zhao, L.: A compact leaky-wave antenna using a planar spoof surface plasmon polariton structure. Int. J. RF Microw. Comput. Aided Eng. **29**, e21617 (2019)
21. Ge, S., Zhang, Q., Chiu, C., Chen, Y., Murch, R.D.: Single-side-scanning surface waveguide leaky-wave antenna using spoof surface plasmon excitation. IEEE Access **6**, 66020–66029 (2018)
22. Zhang, Q., Zhang, Q., Chen, Y.: High-efficiency circularly polarised leaky-wave antenna fed by spoof surface plasmon polaritons. IET Microw. Antennas Propag. **12**(10), 1639–1644
23. Wei, D., Li, J., Yang, J., Qi, Y., Yang, G.: Wide-scanning-angle leaky-wave array antenna based on microstrip SSPPs-TL. IEEE Antennas Wirel. Propag. Lett. **17**(8), 1566–1570 (2018)

24. Wang, C.-J., Jou, C.F., Wu, J.-J.: A novel two-beam scanning active leaky-wave antenna. IEEE Trans. Antennas Propag. **47**(8), 1314–1317 (1999)
25. Hu, C.-C., Jsu, C.F., Wu, J.-J.: An aperture-coupled linear microstrip leaky-wave antenna array with two-dimensional dual-beam scanning capability. IEEE Trans. Antennas Propag. **48**(6), 909–913 (2000)
26. Chen, T.-L., Lin, Y.-D.: Dual-beam microstrip leaky-wave array excited by aperture-coupling method. IEEE Trans. Antennas Propag. **51**(9), 2496–2498 (2003)
27. Li, Y., Xue, Q., Yung, E.K., Long, Y.: Fixed-frequency dual-beam scanning microstrip leaky wave antenna. IEEE Antennas Wirel. Propag. Lett. **6**, 444–446 (2007)
28. Karmokar, D.K., Esselle, K.P., Bird, T.S.: Wideband microstrip leaky-wave antennas with two symmetrical side beams for simultaneous dual-beam scanning. IEEE Trans. Antennas Propag. **64**(4), 1262–1269 (2016)
29. Zhuang, K., et al.: Pattern reconfigurable antenna applying spoof surface plasmon polaritons for wide angle beam steering. IEEE Access **7**, 15444–15451 (2019)
30. Zhang, Y., Han, Z.: Spoof surface plasmon based planar antennas for the realization of terahertz hotspots. Sci. Rep. **5**, 18606 (2015)
31. Yin, J.Y., Ren, J., Zhang, H.C., Pan, B.C., Cui, T.J.: Broadband frequency-selective spoof surface plasmon polaritons on ultrathin metallic structure. Sci. Rep. **5**, Art. no. 8165 (2015)
32. Zhang, X., Fan, J., Chen, J.: High gain and high-efficiency millimeter-wave antenna based on spoof surface plasmon polaritons. IEEE Trans. Antennas Propag. **67**(1), 687–691 (2019)
33. Han, Y., Wang, J., Gong, S., Li, Y., Zhang, Y., Zhang, J.: Low RCS antennas based on dispersion engineering of spoof surface plasmon polaritons. IEEE Trans. Antennas Propag. **66**(12), 7111–7116 (2018)
34. Lv, X., Cao, W., Zeng, Z., Shi, S.: A circularly polarized frequency beam-scanning antenna fed by a microstrip spoof SPP transmission line. IEEE Antennas Wirel. Propag. Lett. **17**(7), 1329–1333 (2018)
35. Liao, D., Zhang, Y., Wang, H.: Wide-angle frequency-controlled beam-scanning antenna fed by standing wave based on the cutoff characteristics of spoof surface plasmon polaritons. IEEE Antennas Wirel. Propag. Lett. **17**(7), 1238–1241 (2018)
36. Tian, D., et al.: Low-profile high-efficiency bidirectional endfire antenna based on spoof surface plasmon polaritons. IEEE Antennas Wirel. Propag. Lett. **17**(5), 837–840 (2018)
37. Chen, H., et al.: Wideband frequency scanning spoof surface plasmon polariton planar antenna based on transmissive phase gradient metasurface. IEEE Antennas Wirel. Propag. Lett. **17**(3), 463–467 (2018)
38. Kandwal, A., Zhang, Q., Tang, X., Liu, L.W., Zhang, G.: Low-profile spoof surface plasmon polaritons traveling-wave antenna for near-endfire radiation. IEEE Antennas Wirel. Propag. Lett. **17**(2), 184–187 (2018)
39. Fan, Y., et al.: Frequency scanning radiation by decoupling spoof surface plasmon polaritons via phase gradient metasurface. IEEE Trans. Antennas Propag. **66**(1), 203–208 (2018)
40. Guan, D., You, P., Zhang, Q., Lu, Z., Yong, S., Xiao, K.: A wide-angle and circularly polarized beam-scanning antenna based on microstrip spoof surface plasmon polariton transmission line. IEEE Antennas Wirel. Propag. Lett. **16**, 2538–2541 (2017)
41. Huang, J.-H., Jou, C.F.: A dual beam scanning microstrip antenna. Prog. Electromagn. Res. Lett. **47**, 125–129 (2014)
42. Anwar, R.S., Ning, H., Mao, L.: Recent advancement in surface plasmon polaritons-plasmonic in sub wavelength structures in microwave and terahertz regimes. Digit. Commun. Netw. **4**, 244–257 (2018)
43. Guo, J., Zhang, H., Wu, S., Liao, Z., Cai, J.: Miniaturized design of multi-octave PA using spoof surface plasmon polaritons. Mater. Des. **242** (2024)
44. Sharma, S., Singh, R.K., Basu, A., Koul, S.K.: Reduction of in-band interfering signals and minimize emission in out-of-band frequencies for EMI/EMC applications. IEEE Trans. Electromagn. Compat. https://doi.org/10.1109/TEMC.2024.3472795
45. Sharma, S., Singh, R.K., Basu, A., Koul, S.K.: Spoof surface plasmon polariton based frequency scanning antennas. IETE J. Res. **69**(11), 7787–7795 (2023)

46. Sharma, S., et al.: A wide-angle forward to backward scanning wideband leaky-wave antenna based on microstrip spoof surface plasmon polaritons. IET Microw. Antennas Propag. **16**(2–3), 91–102 (2022)
47. Sharma, S., Singh, R.K., Basu, A., Koul, S.K.: A wideband transition from microstrip line to microstrip spoof surface plasmon polariton line for microwave/millimeter-wave applications. Int. J. RF Microw. Comput. Aided Eng. 6485834, 8 pp (2023)
48. Sharma, S., Singh, R.K., Basu, A., Koul, S.K.: Spoof surface plasmon polariton transmission lines for EMI reduction. In: 2023 Joint APEMC and INCEMIC, Bangalore, India, pp. 1–3 (2023)
49. Sharma, S., Singh, R.K., Basu, A., Koul, S.K.: Capacitively coupled spoof surface plasmon unit cells based transmission line with band rejection capability. In: IEEE Microwaves, Antennas and Propagation Conference (MAPCON) 2022, Bangalore, India (2022)
50. Sharma, S., Singh, R.K., Basu, A., Koul, S.K.: Leaky wave antenna design by loading radiating elements to an asymmetrical SSPP line. In: The 2022 International Symposium on Antennas and Propagation (ISAP), Darling Harbor, Sydney, Australia, pp. 1–2 (2022)
51. Sharma, S., Singh, R.K., Basu, A., Koul, S.K.: Microstrip spoof surface plasmon polariton based leaky wave antenna. In: IEEE Indian Conference on Antennas and Propagation (InCAP), pp. 897–900 (2021)
52. Sharma, S., Abegaonkar, M.P.: Gain and aperture efficiency improvement using uniplanar EBG superstrate. In: 2018 IEEE Indian Conference on Antennas and Propagation (InCAP), Hyderabad, India, pp. 1–3 (2018)
53. Sharma, S., Singh, R.K., Basu, A., Koul, S.K.: Planar slot antenna array based on spoof surface plasmon transmission line. In: 2023 IEEE Microwaves, Antennas, and Propagation Conference (MAPCON), Ahmedabad, India, pp. 1–4 (2023). https://doi.org/10.1109/MAPCON58678.2023.10463715
54. Sharma, S., Abegaonkar, M.P.: Spoof localized surface plasmon polariton for sensing applications. In: IEEE IMBioC (2025)
55. Singh, R.K., Michel, A., Nepa, P., Salvatore, A.: Wearable dual-band quasi-Yagi antenna for UHF-RFID and 2.4 GHz applications. IEEE J. Radio Freq. Identif. (RFID) **4**(4), 420–427 (2020). https://doi.org/10.1109/JRFID.2020.3000298
56. Singh, R.K., Michel, A., Nepa, P., Salvatore, A., Terraroli, M., Perego, P.: Compact and wearable Yagi-like textile antennas for near-field UHF-RFID readers. IEEE Trans. Antenna Propag. **69**(3), 1324–1333 (2021). https://doi.org/10.1109/TAP.2020.3030944
57. Michel, A., Singh, R.K.: On the use of power-transfer efficiency to analyze the performance of a 3D-printed wearable UHF-RFID antenna. URSI Radio Sci. Bull. **2021**(377), 38–45 (2022). https://doi.org/10.23919/URSIRSB.2021.9829346
58. Michel, A., Singh, R.K., Nepa, P., Manara, G.: A smart glove for near-field UHF RFID applications. URSI Radio Sci. Lett. **3** (2021). https://doi.org/10.46620/21-0046
59. Kannan, S., Singh, R.K., Ray, K.P., Singh, M.: A compact and flexible dipole antenna to be integrated into military berets. In: IEEE Microwaves, Antennas and Propagation Conference (MAPCON) 2022, Bangalore, India (2022)
60. Singh, R.K., Michel, A., Nepa, P., Salvatore, A.: Compact quasi-Yagi reader antenna for UHF RFID smart-glove. In: IEEE International Conference on Smart and Sustainable Technologies, Croatia, September 23–26 (2020)
61. Singh, R.K., Michel, A., Nepa, P., Salvatore, A.: Modified Yagi-Uda reader antenna for UHF RFID smart-glove. In: 14th European Conference on Antennas and Propagation (EUCAP), pp. 1–4, Copenhagen, Denmark (2020)
62. Singh, R.K., Michel, A., Nepa, P., Salvatore, A.: Glove integrated dual-band Yagi reader antenna for UHF RFID and Bluetooth application. In: International Workshop on Antenna Technology (iWAT), pp. 1–4, Bucharest, Romania (2020)
63. Singh, R.K., Michel, A., Nepa, P., Salvatore, A.: Glove integrated solenoid antennas for near-field UHF RFID applications. In: IEEE Indian Conference on Antennas and Propagation (InCAP), pp. 1–4, Ahmedabad, India (2019)

64. Singh, R.K., Michel, A., Nepa, P., Salvatore, A.: Design of a compact Yagi-Uda antenna for near field UHF RFID smart gloves. In: IEEE International Conference on RFID Technology and Applications (RFID-TA), pp. 1–4, Pisa, Italy (2019)

Chapter 2
Transition from Conventional Transmission Lines to SSPP

2.1 Introduction

Transition plays an important role in feeding the signal in SSPP-based structures. It is difficult to feed into and extract RF signals directly from the components or circuits realized using plasmonic technology. An efficient transition, which can convert an existing mode (e.g., quasi-TEM mode of a microstrip line) into an SSPP mode, is needed so that the signals can be extracted from or fed to the circuits efficiently. These conversions from spatial waves to SSPP were obtained using prism and gratings [1]. Nowadays, many antennas, filters, and power dividers are being developed using SSPP technology. Both grounded and ungrounded SSPPs are used to develop microwave components and circuits [2]. Transitions from a conventional transmission line to an ungrounded SSPP line have been reported in [3–8]. Although the microstrip to ungrounded SSPP or CPW to ungrounded SSPP transitions has been realized, grounded SSPP-based transitions find application in realizing passive microwave circuits such as filters where the ground is needed for the realization of short or open-circuited stubs.

Poor grounding degrades the transmission of a signal and leads to unwanted radiation. Also, an amplifier possessing high power gain requires a good grounding to suppress feedback (output to the input) signal to prevent oscillation. The advantage of using grounded SSPP is that it can be easily integrated with the microstrip and CPW technology. As the SSPP technology is maturing [9–17], we need a proper ground to solve the above-mentioned issues. An ultra-thin corrugated metallic strip with mirror symmetry on both sides of the dielectric substrate has been used to integrate the amplifier chip with SSPP [18]. It is beneficial to realize a transition with the ground plane on the back side so that microwave active and passive circuits can easily be integrated. The transition occupation ratio is defined as the ratio of the length of transition to the wavelength which is calculated based on the center frequency of the operating frequency band. The transition occupation ratio defines the compactness of the transition as smaller the value the more compact the transition.

S. K. Koul and S. Sharma, *Microwave Integrated Circuits Based on Spoof Surface Plasmon Polariton*, Signals and Communication Technology,
https://doi.org/10.1007/978-981-96-9041-1_2

The ratio of the receiving end power to the sending end power of a transmission line is known as the transmission efficiency of the line. Each of these variables is crucial to the creation of an effective transition. We will examine various transition types in the microwave frequency range in this chapter's upcoming sections by employing conventional transmission lines as input.

2.2 Transition from Microstrip to SSPP

The transition from a microstrip line to an SSPP transmission line plays a key role in the research of SSPP as a microstrip line is the widely used transmission structure in the microwave domain. These transitions are used to convert quasi-TEM mode in one type of transmission line to quasi-TM mode in another type of transmission line. In [19–21], transitions from conventional microstrip lines to grounded SSPPs have been proposed. In [19], a miniaturized SSPP low-pass filter has been designed with a compact transition where the cut-off frequency of the filter was 14 GHz. A gradient corrugated strip with 'under layer ground' was used to match the impedance of the microstrip with the grounded SSPP, the realized transition had high conversion efficiency, but the proposed transition was large in size, and the transition operated well up to 6.5 GHz [20]. A broadband transition from QTEM mode to SSPP mode has been reported in [21] where it was shown that conversion of the wave vectors could be achieved by gradually varying the groove height. Although that transition has achieved high conversion efficiency, the operating frequency of the transition is limited by the groove height. According to available reports, it is observed that it is difficult to realize a wideband transition while maintaining good performance. To realize a transition at higher frequencies, the groove height becomes very less, and it is difficult to realize such transitions using conventional photolithography or machining processes.

Figure 2.1a shows the geometry of an SSPP unit cell. The width of the groove is 'a' on the upper side and 'b' on the lower side. The periodicity of the unit cell is 'p', and the height of the groove is 'h'. Figure 2.1b–d shows the dispersion relation of a single unit cell by considering different values of 'a' and 'b'. The unit cells used in the transition design for $p = 1$ mm, $p = 2$ mm and $p = 3$ mm are shown as the inset in graphs of Fig. 2.1. Figure 2.1b, c and d shows the effect of variation of 'a' and 'b' on the asymptotic cut-off frequency at $p = 1$ mm and $h = 0.4$ mm, $p = 2$ mm and $h = 0.8$ mm, and $p = 3$ mm and $h = 1.7$ mm, respectively. Figure 2.1 shows the sensitivity or gradual increase in the wave momentum with the change in the shape of a unit cell and this helps in realizing a smooth transition.

Figure 2.2a–c shows the extracted characteristic impedance plots. Characteristic impedances are plotted using the cubic spline interpolation method [22]. Figure 2.2a, b, and c shows the effect of variation of 'a' and 'b' on the characteristic impedance against the frequency for $p = 1$ mm and $h = 0.4$ mm, $p = 2$ mm and $h = 0.8$ mm and $p = 3$ mm and $h = 1.7$ mm, respectively. The transition structure is not only

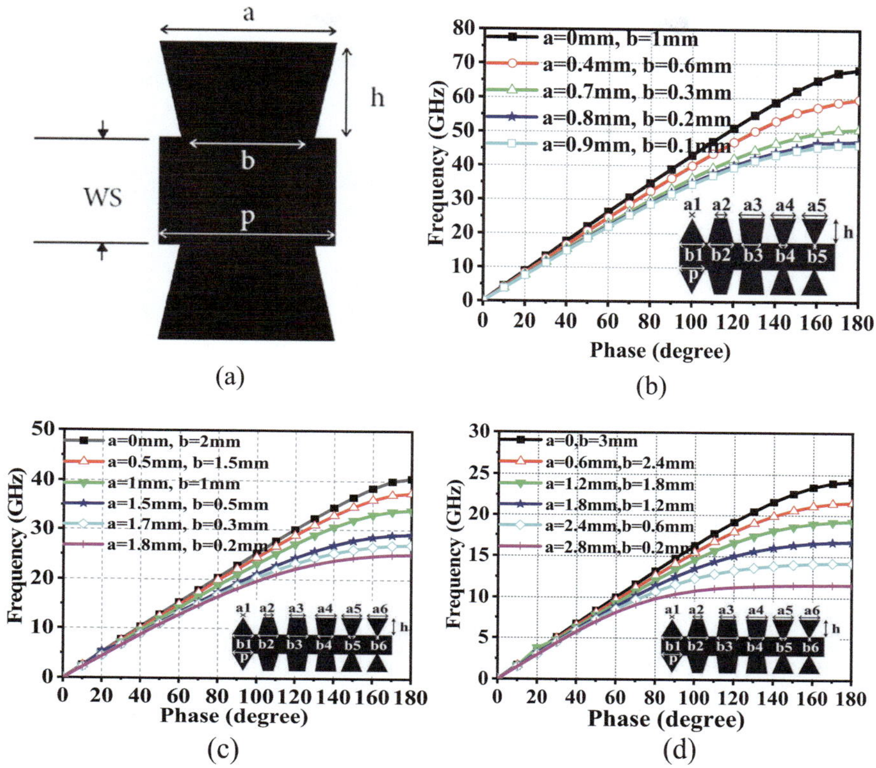

Fig. 2.1 **a** Unit cell, **b–d** dispersion diagrams of unit cell with shape variation ('*a*' and '*b*' variation) for **b** $p = 1$ mm and $h = 0.4$ mm **c** $p = 2$ mm and $h = 0.8$ mm **d** $p = 3$ mm and $h = 1.7$ mm [7]

realized by getting impedance and momentum matching but also by transforming quasi-TEM mode to SSPP mode. Figure 2.3a, b shows the normal ($|E_x|$) and azimuthal ($\sqrt{(|E_z|^2 + |E_y|^2)}$) component of the electric field at different positions on a plane perpendicular to the transition at 10 GHz for $p = 2$ mm.

In Fig. 2.3, plane 1 shows that the normal component of the electric field for the microstrip line is zero and most of the field is concentrated in the azimuthal component. Further, as we move far from the plane perpendicular to the transition, the azimuthal component decreases gradually, and the normal component increases gradually. One can observe that the electric field concentration is higher for the normal component than the azimuthal component in plane 4. The electric field study shows that the field is concentrated on the surface of the metal–dielectric interface. This indicates that the transition efficiently converts a QTEM mode into the SSPP mode.

Fig. 2.2 Characteristic impedances of SSPP-TL for **a** $p = 1$ mm and $h = 0.4$ mm **b** $p = 2$ mm and $h = 0.8$ mm, **c** $p = 3$ mm and $h = 1.7$ mm [7]

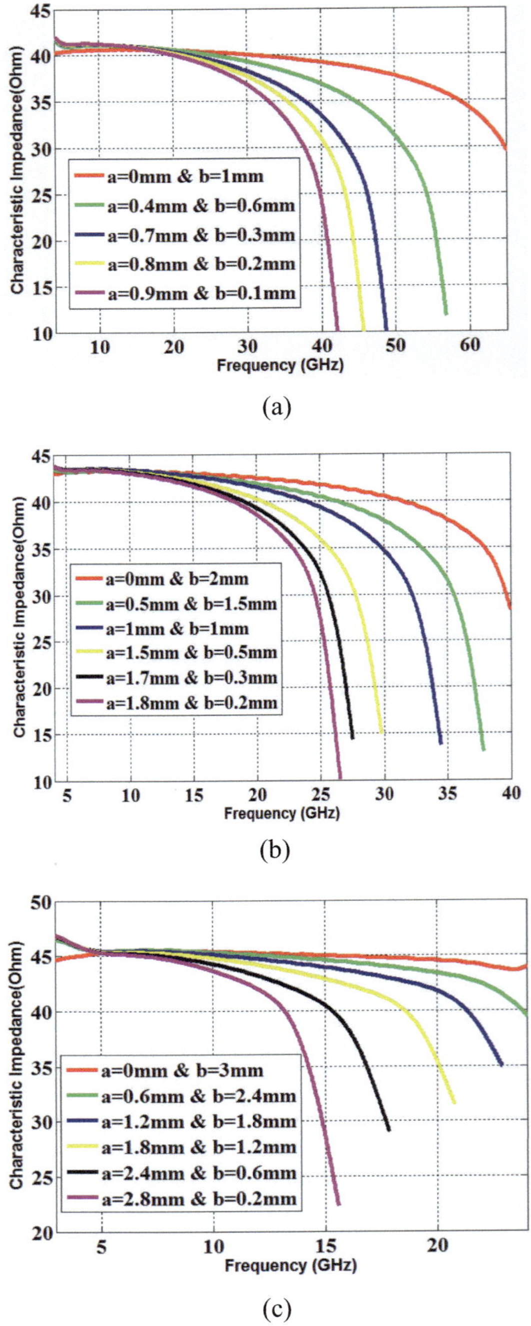

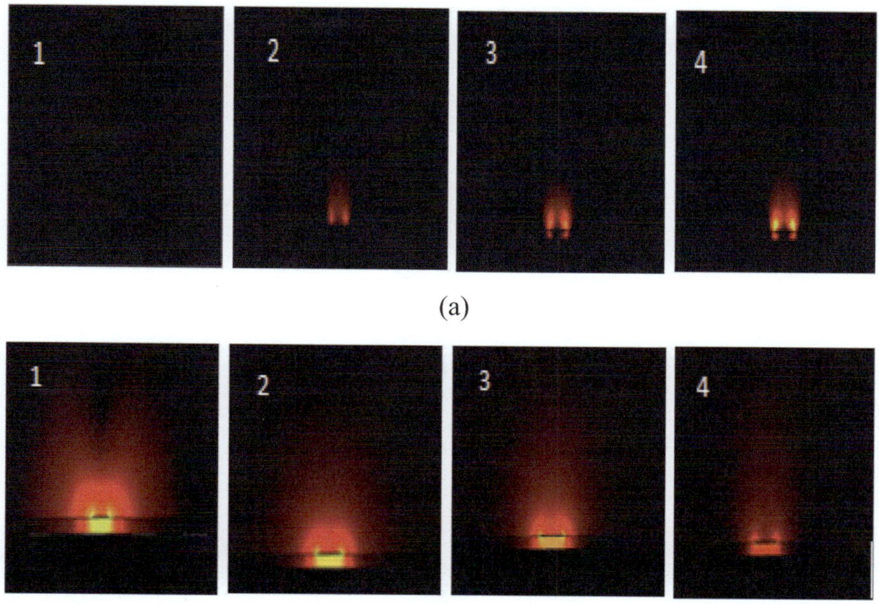

Fig. 2.3 Electric-field distribution for SSPP-TL for $p = 2$ mm **a** normal component **b** azimuthal component [7]

Figure 2.4 shows the SSPP-TL with the proposed transition. It is concluded from the above study as shown in Figs. 2.3, 2.4, and 2.5 that the momentum, impedance, and polarization matching are achieved using this transition. It is designed on an F4B dielectric substrate whose dielectric constant is 2.65, loss tangent is 0.001, and height of substrate is 1.5 mm for $p = 3$ mm and 0.5 mm for $p = 2$ mm. whereas, for $p = 1$ mm, the SSPP-TL is designed on Rogers 4350B with a dielectric constant of 3.66, loss tangent of 0.0037, and height of 0.254 mm. The SSPP-TL has a microstrip line section of length L_1 and width W_s.

The length of the transition and transmission line is L_2 and L_3, respectively. The transition is designed by varying the shape of the unit cell. The parameters 'a' and

Fig. 2.4 SSPP-TL with the proposed transition

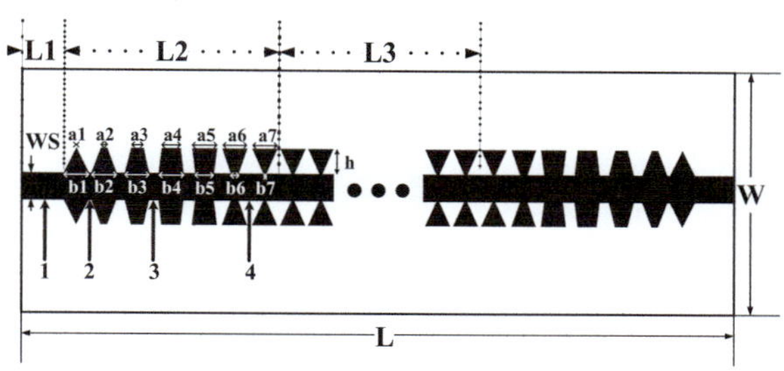

Fig. 2.5 S-parameters of SSPP-TL at different groove heights for **a** $p = 3$ mm, **b** $p = 2$ mm, and **c** $p = 1$ mm [7]

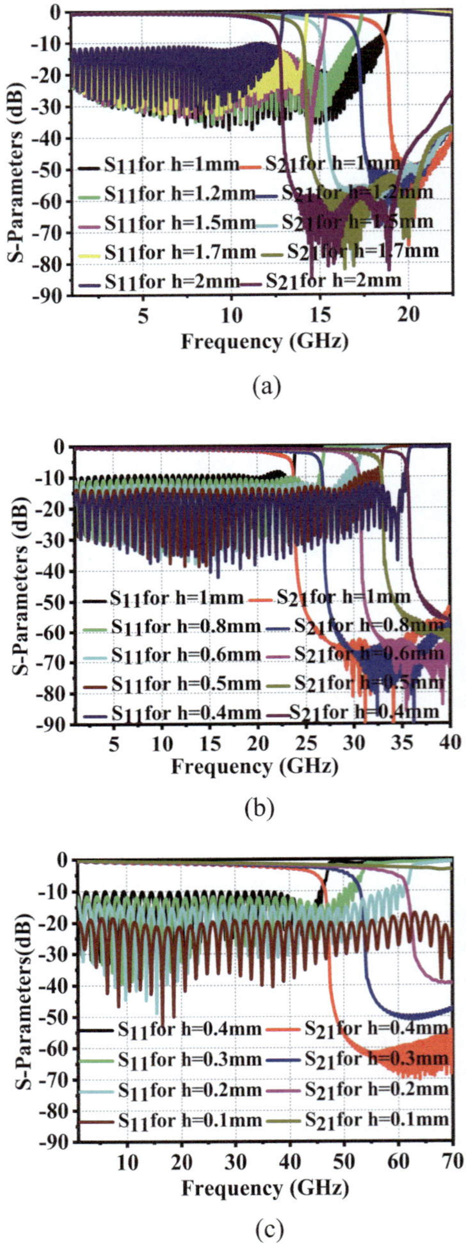

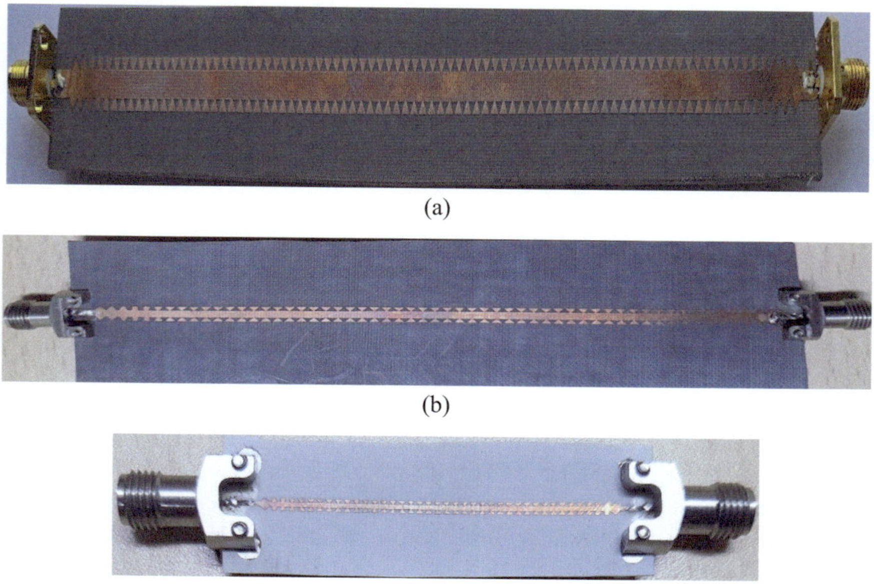

(a)

(b)

(c)

Fig. 2.6 Photographs of fabricated SSPP-TLs with proposed transition for **a** $p = 3$ mm **b** $p = 2$ mm **c** $p = 1$ mm [7]

'b' of the unit cell are varied to realize a smooth transition. For $p = 1$ mm, 'a' is varied from 0 to 0.9 mm while 'b' is varied from 1 to 0.1 mm. The length of the transition, L_2 is 5 mm for $p = 1$ mm. For $p = 2$ mm, 'a' is varied from 0 to 1.8 mm while 'b' is from 2 to 0.2 mm, the length of the transition, L_2 is 12 mm for $p = 2$ mm. For $p = 3$ mm, the variation in 'a' and 'b' is from 0 to 2.8 mm and 3 to 0.2 mm, respectively, and the length of the transition, L_2 is 18 mm.

The transition is compact and robust; one can easily tune the low-pass characteristic of the line up to 40 GHz by varying the geometrical parameter of the unit cell. When the SSPP line is designed with the conventional transition with $p = 1$ mm and $h = 0.4$ mm, the height of the unit cells is varied from 0.04 to 0.4 mm which is not realizable using conventional photolithography or machining processes. The objective of the proposed work was to realize a compact and wideband transition and the calculated dimensions of the transition must be larger than 0.1 mm so that it can be realized using conventional photolithography process.

The S-parameters of the SSPP-TL with the proposed transition are shown in Fig. 2.5a–c. These transmission lines work as low-pass filters whose cut-off frequency can be varied by varying the height and shape of the unit cell. Figure 2.5a–c shows the S-parameters of SSPP-TL by varying the height (h) for $p = 3$ mm, $p = 2$ mm, and $p = 1$ mm, respectively. The graphs depict that the low-pass characteristic can be varied from lower to higher frequencies with good transition occupation ratio and transmission efficiency by varying the geometrical parameters.

Three transmission lines with different cut-off frequencies are designed in a full-wave CST Microwave Studio and demonstrated. Figure 2.6a, b, and c shows the fabricated SSPP-TLs with the proposed transition for $p = 3$ mm, 2 mm, and 1 mm, respectively. Figure 2.7a, b, and c shows the simulated and measured S-parameters of the SSPP-TL for $p = 3$ mm, $p = 2$ mm, and $p = 1$ mm, respectively.

Measured and simulated results in Fig. 2.7a and b show that the reflection coefficient is better than −10 dB and the transmission coefficient is better than −3 dB in the frequency range from 1–12 GHz, 1–22 GHz, for $p = 3$ mm, $p = 2$ mm, respectively. In Fig. 2.7c, the simulated and measured reflection coefficient is better than −10 dB, but the transmission coefficient is lower than −3 dB at the higher frequencies. The insertion loss of the coaxial connector is increased at the higher frequencies. A small discrepancy in simulated and measured results is due to fabrication imperfections and feed connectors that are not considered in the simulation.

In this section, a compact and wideband transition from microstrip line to microstrip SSPP-TL is studied in detail. The transition is different from the conventional transition as the shape of the unit cell is varied instead of height. The advantage of varying only height is the ease of fabrication at higher frequencies. At higher frequencies, the gradient height of the unit cell for transition makes the fabrication complex. The obtained transition occupation ratio and transition efficiency for all three designed cases are excellent. This transition is a potential candidate to be used in circuits designed at microwave and millimeter-wave frequencies.

2.3 Transition from Coplanar Waveguide to SSPP

Owing to its superior isolation over microstrip, coplanar waveguide (CPW) is primarily utilized in microwave devices and MMICs. The fact that the conductors are all above the substrate, allowing for the use of a thick substrate, is another benefit of CPW for MMIC use. In the transition from coplanar waveguide to SSPP transmission line, quasi-TEM is transitioned to quasi-TM mode. In the CPW to ungrounded SSPP transitions which have been realized by Ma et al. [23] and Xu et al. [24], respectively, the flared ground is used to improve the impedance matching and reduce the insertion loss. Tapered corrugation is used to improve impedance, polarization, and momentum matching of the transition.

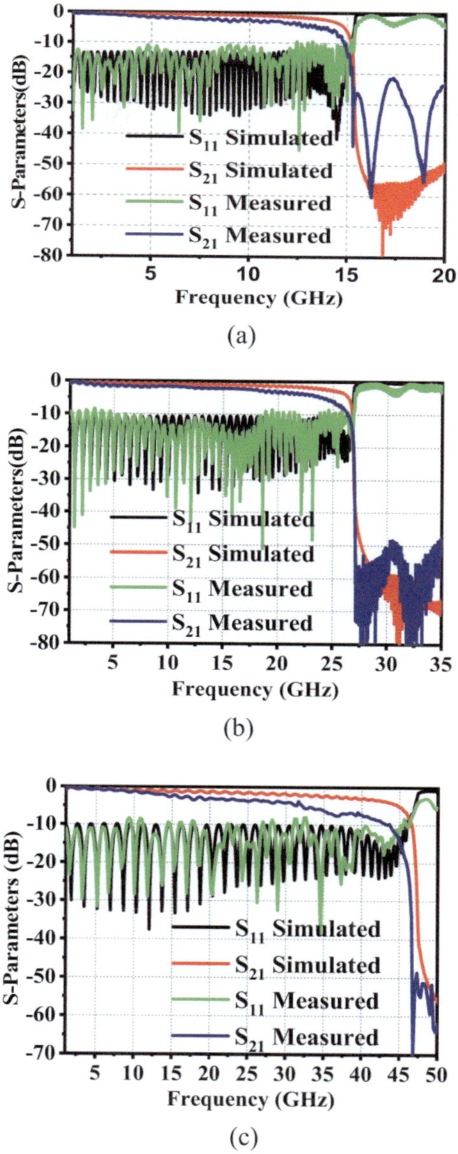

Fig. 2.7 Simulated and measured S-parameters of SSPP-TL for **a** $p = 3$ mm and $h = 1.7$ mm **b** $p = 2$ mm and $h = 0.8$ mm **c** $p = 1$ mm and $h = 0.4$ mm [7]

2.4 Transition from Slot Line to SSPP

In the slot line, the propagating mode is quasi-transverse electric (TE) whereas in the SSPP transmission line, the propagating mode is quasi- transverse magnetic (TM). We require an efficient transition to convert quasi-TE to quasi-TM which is possible when an electric component is generated in the longitudinal direction to transmit wave from the slot line to the SSPP transmission line. A radial stub is used with both the microstrip line and the slot line for wideband impedance matching. This transmission line finds application with ferrites and the circuits where shunt-mounted elements are used. In [25], a brief segment of the tapered slot line that connects to the SSPPs at an inclination angle serves as both an impedance and a waveform transformer, allowing for the successful realization of a compact configuration. To further minimize the insertion loss, a second metallic ground in the shape of a trapezoid has been used.

A slot line SSPP transmission line can be designed using a corrugated slot line and a group of additional interdigital structures (ISs) which adds an extra parameter to control cut-of property [26]. Figure 2.8a shows the schematic whereas (b) shows the fabricated structure of the slot line SSPP transmission line with interdigital structures. The number of interdigital structures contributes to the transmission cut-off frequency as the number of ISs increases the transmission cut-off frequency decreases. Since the width of the CSSPP waveguide is directly related to the operating frequency, the degree of miniaturization can be modulated freely by carefully choosing the number of the ISs. Figure 2.9 shows the S-parameter of the slot line-based SSPP.

2.5 Conclusion

In this chapter, the transitions from microstrip line to grounded SSPP have been studied. Different parameters which define the transition characteristics are defined in this chapter. We have studied different transitions from conventional transmission lines such as microstrip lines, coplanar waveguides, and slot lines to SSPP transmission lines. Advantages and disadvantages of the grounded and groundless SSPP structures are reported in this chapter. We studied a different transmission technique which helped to resolve the problem of fabricating a transmission line at higher frequency range. This chapter provides an insight into the unit cell design for different transition structures and characteristics to design an efficient transition.

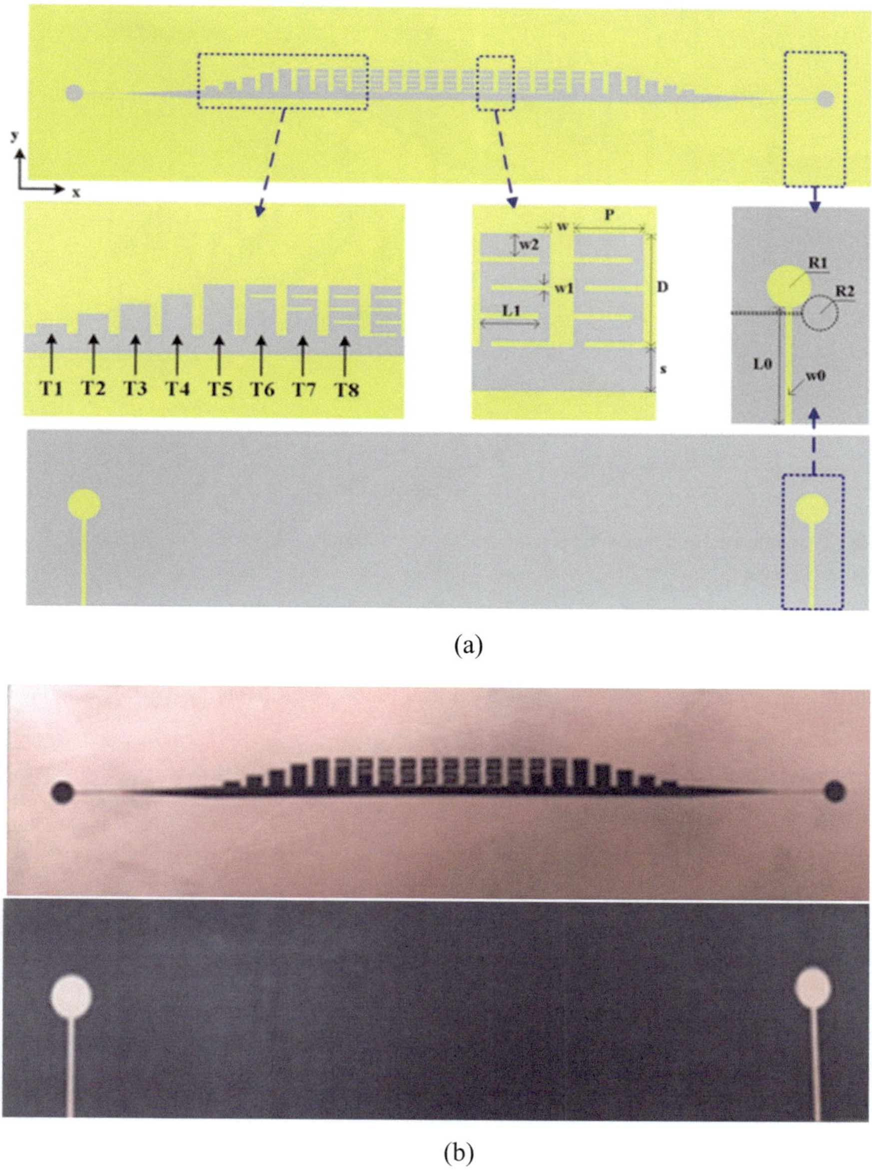

Fig. 2.8 **a** Schematic and **b** fabricated structure of the proposed design of slot SSPP transmission line [23]

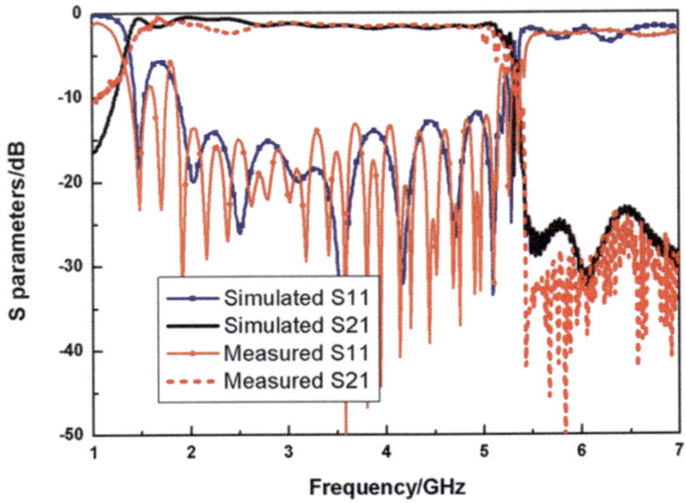

Fig. 2.9 Simulated and measured S-parameter of the slot line based of SSPP technology [23]

References

1. Raether, H.: Surface Plasmons on Smooth and Rough Surfaces and on Gratings. Springer, Berlin (1988)
2. Han, Y., Wang, J., Li, Y., Gong, S., Zhang, J., Qu, S.: A frequency scanning antenna based on hybridization of the quasi-TEM mode and spoof surface plasmon polaritons mode. J. Phys. D Appl. Phys. **52**(38), 38LT01 (2019)
3. Cao, D., Li, Y., Wang, J.: Wideband compact slotline-to-spoof-surface plasmon-polaritons transition for millimeter waves. IEEE Antennas Wirel. Propag. Lett. **16**, 3143–3146 (2017)
4. Yan, R.T., et al.: A broadband and high-efficiency compact transition from microstrip line to spoof surface plasmon polaritons. IEEE Microw. Wirel. Compon. Lett. **30**(1), 23–26 (2020)
5. Kianinejad, A., Chen, Z.N., Qiu, C.: Design and modeling of spoof surface plasmon modes-based microwave slow-wave transmission line. IEEE Trans. Microw. Theory Tech. **63**(6), 1817–1825 (2015)
6. Liao, Z., Zhao, J., Pan, B.C., Shen, X., Cui, T.J.: Broadband transition between microstrip line and conformal surface plasmon waveguide. J. Phys D: Appl Phys. (2014)
7. Sharma, S., Singh, R.K., Basu, A., Koul, S.K.: A wideband transition from microstrip line to microstrip spoof surface plasmon polariton line for microwave/millimeter-wave applications. Int. J. RF Microw. Comput. Aided Eng. Article ID 6485834, 8 pp (2023)
8. Wan, X., Yin, J.Y., Zhang, H.C., Cui, T.J.: Dynamic excitation of spoof surface plasmon polaritons. Appl. Phys. Lett. **105**(8), 083502 (2014)
9. Wei, Y., Wu, Y., Wang, W., Pan, L., Yang, Y., Liu, Y.: Double-sided spoof surface plasmon polaritons-line bandpass filter with excellent dual-band filtering and wide upper band suppressions. IEEE Trans. Plasma Sci. **48**(12), 4134–4143 (2020)
10. Wu, Y., et al.: A new self-packaged substrate integrated air-filled spoof surface plasmon polaritons line with inherent low loss and deep upper frequency suppression. IEEE Trans. Plasma Sci. **48**(10), 3516–3523 (2020)
11. Pan, L., Wu, Y., Wang, W., Wei, Y., Yang, Y.: A flexible high-selectivity single-layer coplanar waveguide bandpass filter using interdigital spoof surface plasmon polaritons of bow-tie cells. IEEE Trans. Plasma Sci. **48**(10), 3582–3588 (2020)

12. Li, M., Wu, Y., Qu, M., Li, Q., Liu, Y.: A novel power divider with ultra-wideband harmonics suppression based on double-sided parallel spoof surface plasmon polaritons transmission line. Int. J. RF Microw. Comput. Aided Eng. **28**, e21231 (2018)
13. Liu, H., Liu, T., Zhang, Q., Ren, B., Wen, P.: Compact balanced bandpass filter design using asymmetric SIR pairs and spoof surface plasmon polariton feeding structure. IEEE Microw. Wirel. Compon. Lett. **28**(11), 987–989 (2018)
14. Guan, D.-F., You, P., Zhang, Q., Yang, Z.-B., Liu, H., Yong, S.-W.: Slow-wave half-mode substrate integrated waveguide using spoof surface plasmon polariton structure. IEEE Trans. Microw. Theory Tech. **66**(6), 2946–2952 (2018)
15. Sharma, S., Singh, R.K., Basu, A., Koul, S.K.: Reduction of in-band interfering signals and minimize emission in out-of-band frequencies for EMI/EMC applications. IEEE Trans. Electromag. Compat. https://doi.org/10.1109/TEMC.2024.3472795
16. Sharma, S., Singh, R.K., Basu, A., Koul, S.K.: Spoof surface plasmon polariton based frequency scanning antennas. IETE J. Res. **69**(11), 7787–7795 (2023). https://doi.org/10.1080/03772063.2023.2252399 (Q3)
17. Sharma, S., et al.: A wide-angle forward to backward scanning wideband leaky-wave antenna based on microstrip spoof surface plasmon polaritons. IET Microw. Antennas Propag. **16**(2–3), 91–102 (2022). https://doi.org/10.1049/mia2.12213 (Q2)
18. Zhang, H.C., Liu, S., Shen, X., Chen, L.H., Li, L., Cui, T.J.: Broadband amplification of spoof surface plasmon polaritons at microwave frequencies. Laser Photon. Rev. **9**(1), 83–90 (2014)
19. Wang, L., Cui, X., Yang, H., Du, Z., Zhao, Y.: Miniaturized spoof surface plasmon polaritons low-pass filter with a novel transition structure. IEEE Photonics Technol. Lett. **31**(15), 1273–1276 (2019)
20. Zhang, W., Zhu, G., Sun, L., Lin, F.: Trapping of surface plasmon wave through gradient corrugated strip with under layer ground and manipulating its propagation. Appl. Phys. Lett. **106**, 021104 (2015)
21. Jaiswal, R.K., Pandit, N., Pathak, N.P.: A novel transition device and multiple band-pass filter using ring resonator based on spoof surface plasmon polaritons at microwave frequency. In: 2017 IEEE MTT-S International Microwave and RF Conference (IMaRC), pp. 1–4 (2017)
22. He, P.H., et al.: Characteristic impedance extraction of spoof surface plasmon polariton waveguides. J. Phys. D: Appl. Phys. **54**, 385102 (2021)
23. Ma, H.F., Shen, X., Cheng, Q., Jiang, W.X., Cui, T.J.: Broadband and high-efficiency conversion from guided waves to spoof surface plasmon polaritons. Laser Photon. Rev. (2014)
24. Xu, J., Cui, Y., Guo, J., Xu, Z., Qian, C., Li, W.: Broadband transition between microstrip line and spoof SP waveguide. Electron. Lett. **52**, 1694–1695 (2016)
25. Cao, D., Li, Y., Wang, J.: Wideband compact slotline-to-spoof-surface plasmon-polaritons transition for millimeter waves. IEEE Antennas Wirel. Propag. Lett. **16**, 3143–3146 (2017). https://doi.org/10.1109/LAWP.2017.2765700
26. Pan, B.C., Luo, G.Q., Liao, Z., et al.: Wideband miniaturized design of complementary spoof surface plasmon polaritons waveguide based on interdigital structures. Sci. Rep. **10**, 3258 (2020)

Chapter 3
Filter Design Using Spoof Surface Plasmon Polariton

3.1 Introduction

A filter is a part of an electronic system that attenuates unwanted signals within the intended operating range and can pass or reject certain frequencies depending on the filter type. The filters in the higher frequency range, i.e., above the microwave range, are designed using the distributed element method. Filters are used in many applications such as selectivity of radio channels, band limiting of noise, and multiplexing of many signals into one channel. Filters are widely used in applications such as microwave links used by fixed communication networks such as television broadcasters and telephone companies. Presently, filters can be found in several mass-produced consumer items such as converters used with satellite television dishes. The filters can be classified as low pass, high pass, bandpass, band-stop (band rejection; notch), or all pass.

SSPP structures offer advantageous option for filter design due to their superior electromagnetic constraint, reduced coupling, and reduced signal crosstalk when compared to microstrip lines [1, 2]. Planar filters with SSPP technology can have small thickness. Because of their innate low-pass transmission characteristics, SSPPs have recently been used to develop a variety of microwave wideband and ultra-wideband (UWB) filters [3–6]. Because of their built-in upper cut-off frequency, SSPPs TL functions as a low-pass filter by itself. Thus, a lower cut-off frequency can be added to create a bandpass filter (BPF) based on SSPPs.

With the quick advancement of microwave technology and the rise in communication frequency, filters are used extensively in many electromagnetic systems. This has led to the publication of numerous studies concentrating on filter properties based on SSPPs. Electric field-coupled lamp (ELC) and metamaterial particle filtering properties were used to design a frequency rejection system [7]. Aperiodic SSPPs TL excited by a coplanar waveguide (CPW) with spiral-shaped defected ground structure (DGS) was used to propose a highly selective wideband BPF [8]. In order to create a BPF that inherited both the upper- and lower cut-off frequencies

S. K. Koul and S. Sharma, *Microwave Integrated Circuits Based on Spoof Surface Plasmon Polariton*, Signals and Communication Technology,
https://doi.org/10.1007/978-981-96-9041-1_3

of the SSPPs and the substrate-integrated waveguide (SIW), a two-conductor SSPPs TL was coupled with SIW [9].

Terahertz is an emerging field with growing interest in the field of 6G/7G communication systems. Filters are important operating devices requires in different imaging, sensor networks, and communication systems. Terahertz-integrated chips which have the potential to be used in imaging, wireless sensing, and high-speed communication have been developing rapidly in recent years and are the key elements for future electronic science and technology. As a vital component of integrated functional chips, on-chip filters play a key role in integrated systems' frequency selection, noise reduction, and clutter suppression.

3.2 Bandpass Filter

A system of RF receivers is made up of several parts, including an oscillator, mixer, bandpass filter (BPF), and antenna. One of the crucial components of a communication transceiver system is the BPF, and the BPF's performance has a big impact on the system's overall performance. Therefore, there is constant interest in the design and development of compact and low-loss planar filters. The design of conventional planar filters typically uses microstrip technology. However, higher ohmic losses and increased crosstalk lead to interference and problems with signal integrity using conventional planar microstrip technology. When compared to microstrip technology, the concept of spoof surface plasmon polaritons (SSPPs) has proven useful in the design of circuits and systems for guiding electromagnetic waves with lower loss, reduced crosstalk, and reduced mutual coupling.

High selectivity, a simple single-layer design, and superior filtering performance are just a few of the benefits of the bandpass filter, which can significantly lower costs and increase communication system efficiency. Unfortunately, a single layer, simple structure, and high selectivity cannot all be obtained in most filters at the same time. Thus, there is an urgent need to design a bandpass filter based on a novel transmission structure. A spoof surface plasmon polaritons (SSPP) line was created as a result. The SSPP-based circuit's planar feature makes it simple to integrate with other current systems and circuits. An integral part of wireless transceiver systems is the bandpass filter (BPF). The design of broadband subsystems, like BPF, is necessary for many broadband applications.

3.3 Techniques to Design Bandpass Filter

The SSPP transmission line has inherent low-pass transmission characteristics so a bandpass filter can be designed by incorporating techniques having high-pass characteristics. So, in this section, we will study different techniques used to design bandpass filters using SSPP. The higher-order mode of the miniaturized SSPP unit cell is used to design a bandpass filter (BPF). The miniaturized unit cell is designed by adding a branch structure at the open end of the stub and bending the waveguide structure at the short end of the stub. The stub has an embedded rectangular slot. The filter's upper cut-off frequency and lower cut-off frequency can be adjusted, respectively, by adjusting the slot's length and width. The cell's upper cut-off frequency can also be independently adjusted by loading various capacitors in between the stubs. The filter has an excellent rectangular coefficient, a center frequency of 8 GHz, and a relative bandwidth of roughly 50% [10].

In [11], the higher-order mode of the spoof surface plasmon polariton (SSPP) unit cell is used to have an excellent bandpass propagation feature. Two SSPP waveguides based on the method with different bandwidths are fabricated and measured to verify the viability of applying the higher-order mode of SSPPs in the design of bandpass filters. The waveguide design process does not include a mode conversion transition section, which prevents additional layout area and lowers device complexity. The proposed high-order SSPP waveguide allows the lower and upper cut-off frequencies of the passband to be independently adjusted by the slot width and length, respectively, in the BPF design. Figure 3.1a, b shows the photograph of the bandpass filter with different slot dimensions, whereas Fig. 3.1c, d shows measured and simulated S-parameters, respectively.

To design a narrow band bandpass filter with a controllable passband, good rejection, and a compact structure, a two-end-extending SSPP transmission line and a pair of L-shaped coupling branches are used. By varying the length of the SSPP transmission line behind the coupling branches and the depth of the SSPP unit, the bandwidth and central frequency can be changed. SSPP bandpass filter achieves a wide 3-dB fractional bandwidth of 11.2% centered at 6.25 GHz and an out-of-band rejection below −15 dB within 1–10 GHz [12]. A miniaturized suspended strip line (SSL) filter based on spoof surface plasmon polariton (SSPP) is described in [13]. Periodic corrugated metallic strips (PCMS) are placed in a rectangular metallic cavity to avoid high-density lossy dielectric fillings. Due to the high dielectric constant, low insertion loss, and light weight of the PCMS structure, miniaturized bandpass filters can be implemented using such a configuration. The bandpass filter operates in the frequency range of 2.7–4.6 GHz with insertion loss as low as 0.1–0.3 dB and the cross-section of the cavity reduced by more than 75%.

In [14], bandpass filter is designed using trident-shaped complementary grooves which have a vertical size reduced by 40% compared to the conventional rectangular groove. The bandpass filter has a controllable frequency band and wide stopband suppression. It has good filtering performance in the frequency band from 2.1 to 8 GHz. A new unit cell is created by modifying the traditional symmetric H-shaped unit cell. A new SSPP structure that offers broadband bandpass filter (BPF) characteristics is designed using this modified unit cell. According to the measurement, BPF has an insertion loss of less than 2 dB and a passband of 1.2–6.0 GHz. The manufactured SSPP BPF measures 26 cm by 4 cm [15]. Figure 3.2a shows the fabricated prototype of the BPF and (b) shows the simulated and measured S-parameter of SSPP. A non-periodic spoof surface plasmon polariton (SSPP) with H-shaped cells is used to design a wideband bandpass filter by integrating the spiral-shaped defected

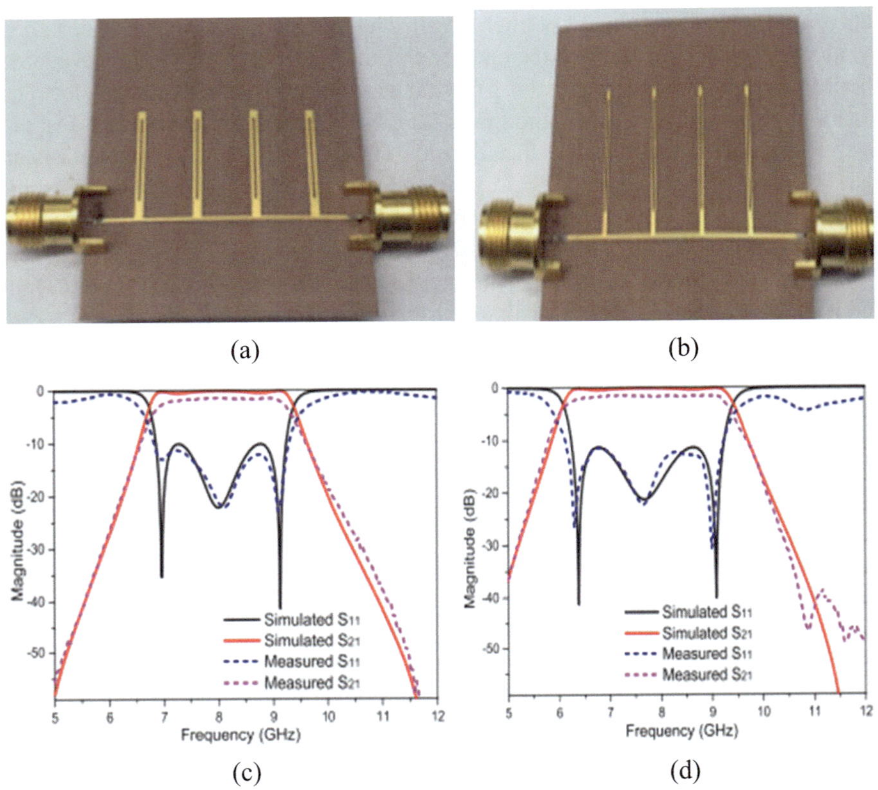

Fig. 3.1 **a**, **b** Bandpass filter using SSPP with different slot dimensions **c**, **d** measured and simulated S-parameter [11]. Reproduced with permission from the IEEE

ground surface (DGS) etched on CPW. The filter shows high passband selectivity ($\Delta f3\text{dB}/\Delta f20\text{dB} = 0.91$) and wide upper stopband with -20 dB rejection [8]. To design a compact single-conductor SSPP transmission line with bandpass characteristics, cross-shaped resonators (CSRs) are integrated into the spoof surface plasmon polaritons (SSPPs) transmission line. Theoretical analysis is carried out to verify the designed structure.

The passband operates in the frequency range from 7.9 to 12.2 GHz with a relative bandwidth of 42%. It finds application in highly integrated SSPP circuits [16]. Double-layered spoof surface plasmon polaritons (DLSSPPs) are used as the main transmission line (MTL) in the design of a broad dual-band bandpass filter. Certain spaces are then loaded onto the MTL by combined-stub dual-mode resonators (CSDMRs). An open stub and a short stub connected in parallel make up each CSDMR. Gradient double-layered metal gratings and the sector structure impedance matching technique are used to achieve efficient mode conversion between the quasi-transverse electromagnetic waves in the microstrip line and the spoof surface plasmon polaritons (SSPPs) on the double-layered surface plasmon waveguide. A parametric

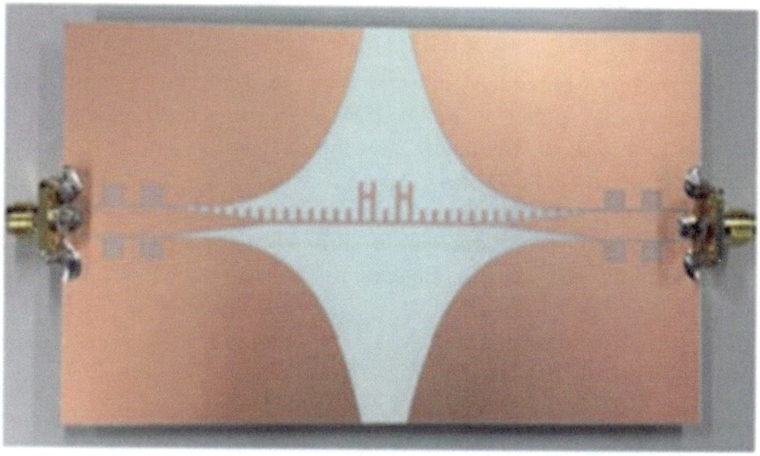

(a)

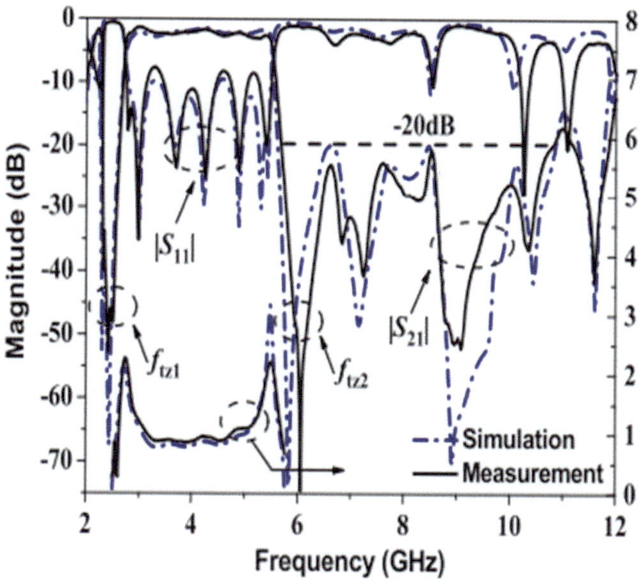

(b)

Fig. 3.2 **a** Photograph of fabricated wideband bandpass filter **b** simulated and measured results of S-parameters and group delay [8] study of the CSDMR shows that by adjusting the stub lengths and widths of the CSDMRs, center frequencies and bandwidths can be controlled. Figure 3.3 shows that the passband loss is reduced in comparison to filters that use single-layered SSPPs. The slow wave effect of the DLSSPPs shortens the distance between two CSDMRs. From the cut-off frequency above, the DLSSPPs low-pass performance shows strong rejection [17] (Fig. 3.4).

A two-sided spoof surface plasmon polaritons-line dual-band bandpass filter is designed using a dual-band filtering section and two SSPP low-pass filtering sections. The suggested dual-band bandpass filter exhibits broad upper band suppressions and

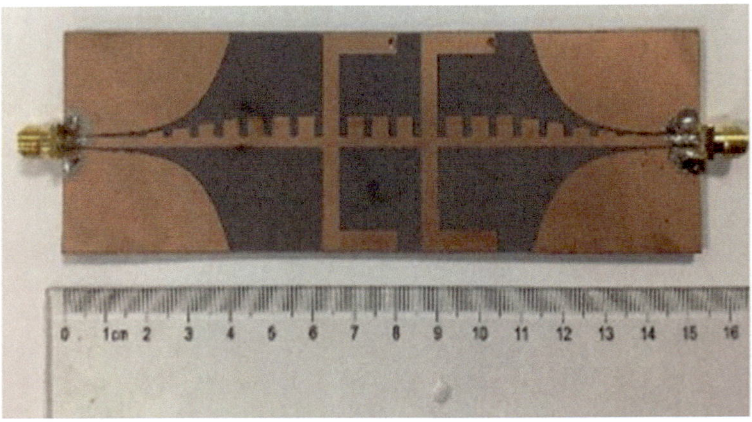

(a)

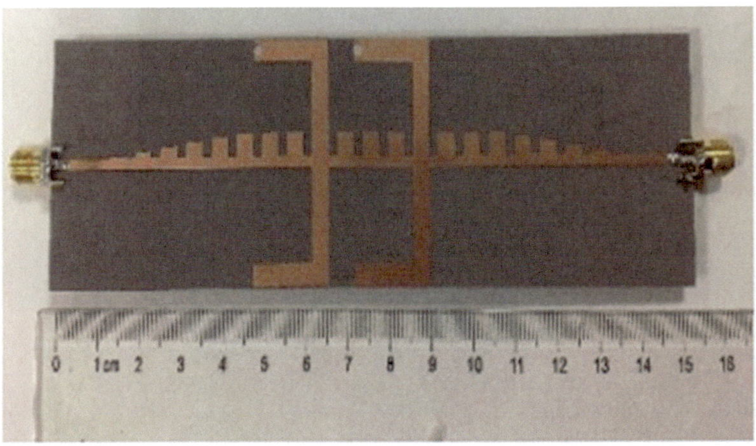

(b)

Fig. 3.3 Photograph of dual-band filter **a** top view **b** bottom view [17]

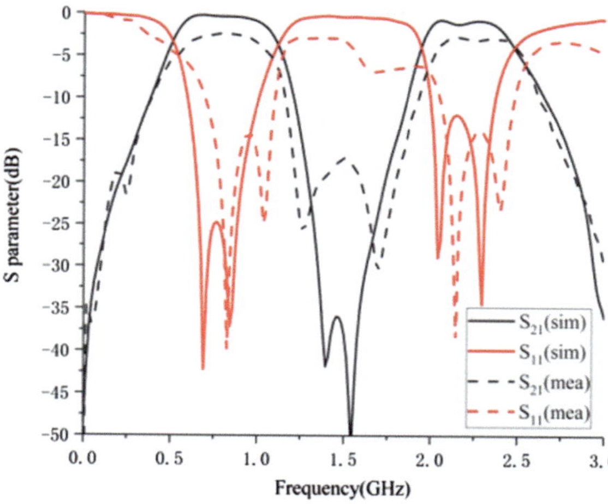

Fig. 3.4 Simulated and measured S-parameters of broad dual-bandpass filter [17]

outstanding dual-band filtering performance. There are two passbands with bandwidths of 41% (2.36–3.58 GHz) and 13.9% (6.7–7.7 GHz). From 9.6 to 39.7 GHz, it can achieve S_{21} suppression below −40 dB [18]. A miniaturized bandpass filter is designed using two double-layer asymmetric SSPPs with fishbone-shaped metal grating, interdigital strip, and meander-line.

To lower the asymptotic frequency in this SSPP structure, a layer of metal gratings and an interdigital strip and meander-line layer are built to increase the effective capacitance. With an exceptionally broad relative bandwidth of 177.3% and the smallest insertion loss of 0.5 dB from 0.29 to 4.82 GHz, the suggested UWB SSPP BPF performs exceptionally well [19]. A wide bandpass broadband bandpass filter in the microwave frequency band is proposed in [20]. It consists of two asymmetrically broken corrugated strips coupled via the grooved strips with an embedded split ring. The proposed bandpass filter has a good band-stop characteristic from 3 to 7 GHz and excellent low-loss performance from 7 to 10 GHz, where its insertion loss is approximately 1.5 dB in the same frequency band. Figure 3.5a shows the photograph of the BPF whereas (b) and (c) shows the transmission and reflection coefficient of the filter. To evaluate the proposed broadband SSPPs filter, a straightforward but precise transmission line model was suggested. An incredibly thin and small waveguide bandpass filter based on SSPPs that has metals covered on the side and bottom surfaces of the substrate in addition to metals etched on the top surface. The width of the rectangular waveguide can be used to adjust the lower cut-off frequency of this filter, while the length of the complementary comb-like structure can be used to adjust the upper cut-off frequency. This filter can inherit the benefits of both the SSPP structures and the rectangular waveguide [21].

It is suggested to use a new double-row staggered double-sided quasi-SSPPs transmission line to create two wideband bandpass filters. A transmission zero can

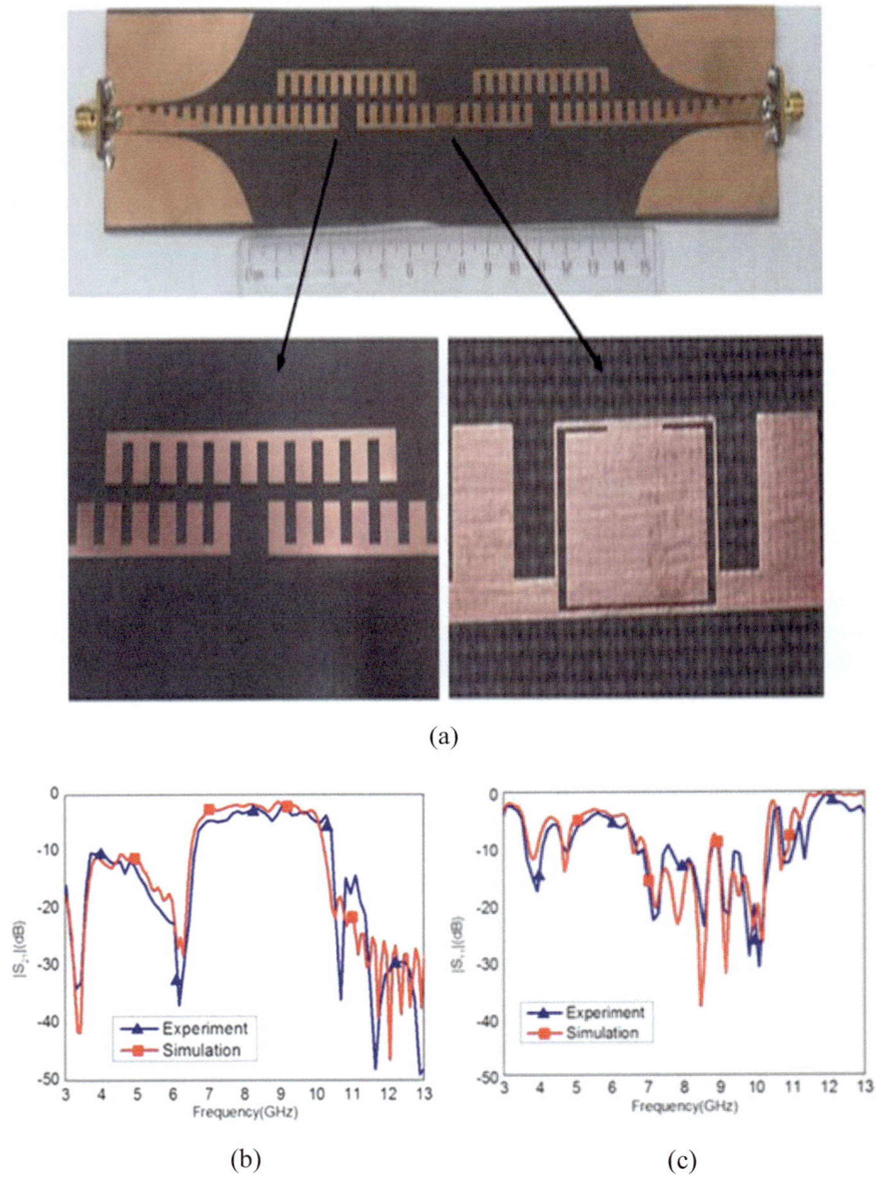

Fig. 3.5 **a** Photograph of the BPF **b** the transmission coefficient **c** the reflection coefficient [20]

be introduced into the original operating frequency band by symmetrically loading U-shaped cells. This creates a wideband bandpass filter with high out-of-band suppression. A double-sided parallel-strip line can be compared to the structure, and to help with analysis and design, a microstrip line equivalent circuit model can be built. An additional improvement in passband selectivity can be achieved by increasing the number of U-shaped cells from three to five. At 11.4 GHz, the measured 3-dB fractional bandwidth results are 18.9%, and the in-band return loss is higher than 10.2 dB. Furthermore, the suppression of out-of-band signals exceeds 20 dB in the lower stopband (from 6.18 to 10.12 GHz) and the upper stopband (from 12.78 to 19 GHz) [22].

A compact, highly efficient bandpass filter with independent tuning for the upper and lower cut-off frequencies is based on coplanar waveguide technology (CPW). The higher cut-off frequency is tuned by a quasi-spoof surface plasmon polaritons (Q-SSPPs) structure, while the low-frequency wave is filtered by the interdigital structure. The suggested Q-SSPPs structure retains the characteristics of the periodic SSPP structures while having just one cross-shaped element. The findings shown in Fig. 3.6 suggest that the design can achieve high transmission efficiency ($|S_{11}| < -15$ dB and $|S_{21}| > -0.2$ dB) and a broad bandwidth from 8.8 to 17 GHz (roughly 63.6%) [23]. A compact, highly efficient THz ultra-broadband bandpass filter that operates between 0.65 and 2.02 THz using spoof surface plasmon polaritons (SSPPs) and coplanar waveguide technology is reported in the literature. Additionally, by adjusting the corresponding parameters, the filter's lower and upper cut-off frequencies can be independently controlled. The low-frequency wave is filtered by the interdigital structure, and the upper cut-off frequency is produced by the SSPPs. Field distributions and dispersion curves explain the operating principles. The suggested filter has an insertion loss of less than 2 dB and a return loss of greater than 11 dB in the passband, respectively [24].

Bow-tie cell interdigital spoof surface plasmon polaritons (SSPP) are used to create a flexible, highly selective single-layer coplanar waveguide (CPW) bandpass filter. One feature of the bandpass filter presented is the independent adjustment of the high and low cut-off frequencies. Bow-tie SSPP cells are utilized to modify the high cut-off frequency, and its interdigital coupling structure is intended to modify the lower one. With excellent selectivity, the filter exhibits a theoretical 3-dB bandwidth from 2.28 to 5.12 GHz (about 76.8%) and -0.4 dB $> |S_{21}| > -1.1$ dB in the passband as shown in Fig. 3.7. This filter can be tested in bent, folded, or even twisted states because it is built using a flexible printed circuit (FPC) board [25].

A 180° phase inverter, two double-layer surface plasmon polaritons (SSPPs) using double-sided parallel-strip line (DSPSL), and two single-layer SSPPs make up a novel differential bandpass filter based on periodic spoof SPPs. The differential-mode and common-mode methods are used to analyze the transmission characteristics of the proposed SSPPs' structure and to design the differential filter. The simulated and measured results show that the differential filter can achieve a wide 20-dB common-mode suppression from 2.9 to 8 GHz, with a 3-dB fractional bandwidth of 25.5% (frequency 5.5 GHz) [26].

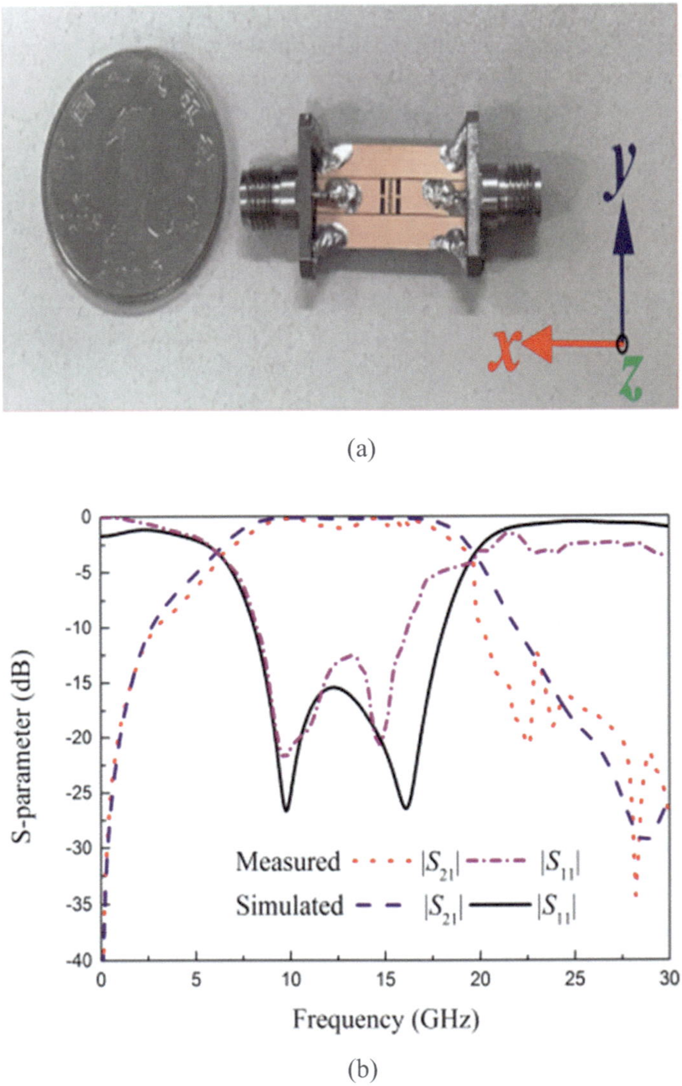

Fig. 3.6 **a** Fabricated SSPP BPF **b** measured and simulated S-parameters [23]

To create a bandpass filter (BPF) that operates at the W-band, a hybrid circuit utilizing double-grating rectangular waveguide (DG-RWG) and spoof surface plasmon polaritons (SSPPs) is suggested. A hybrid SSPP-RWG is created when the suggested SSPPs, which are made of corrugated metallic blocks, are embedded into the RWG. By adjusting their geometric parameters, the lower cut-off frequency of DG-RWG and the upper cut-off frequency of SSPP-RWG can be independently controlled. For this reason, the hybrid circuit made up of DG-RWG and SSPP-RWG

(a)

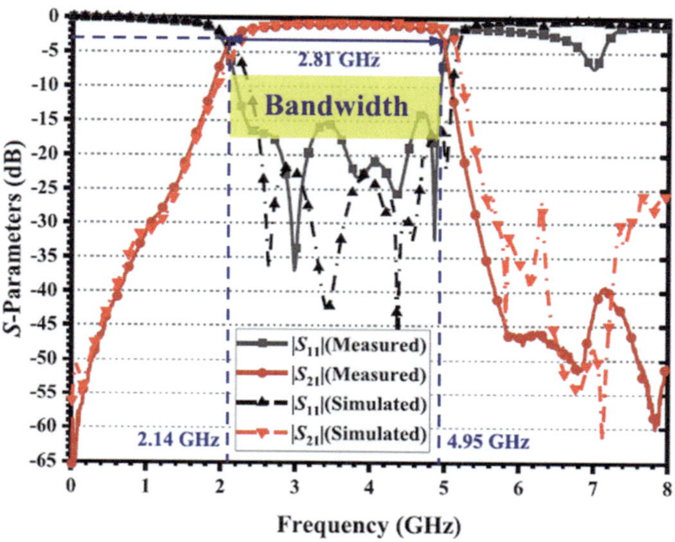

(b)

Fig. 3.7 **a** Fabricated flexible SSPP-based BPF **b** measured and simulated S-parameters [25]. Reproduced with permission from the IEEE

is appealing for designing a BPF with tunable bandwidth. The RWG BPF in W-band provides a low insertion loss of roughly 0.57 dB at the center frequency and operates in the 80.1–93.8 GHz range as shown in Fig. 3.18 [27]. In a groove gap waveguide (GGW), spoof surface plasmon polariton (SSPP) is developed to realize an adjustable bandpass filter for Ka-band applications. The structure is made up of a single row of metallic blocks that have been corrugated and inserted into the GGW groove. The SSPP unit cell's height or period determines the passband's upper cut-off frequency, while the groove width and pin height can be changed to modify the lower cut-off frequency. The filter offers low insertion loss of 0.22 dB over a band from 29.8 to 35.8 GHz [28].

In the microwave range, a novel planar waveguide based on spoof surface plasmon polaritons (SSPPs) using a fishbone corrugated slot structure is designed. With such a structure and tight field confinement of SSPPs, a low-dispersion band can be realized, leading to size miniaturization of the suggested waveguide. Using this proposed waveguide, the high-frequency stopband of the ultra-wideband bandpass filter (BPF) is formed, and the low-frequency stopband is suitably designed by introducing the microstrip-to-slot line transition. From 2.1 to 8 GHz, the reflection coefficient is better than −10 dB, with lowest insertion loss of 0.37 dB at 4.9 GHz [29]. Based on double-layer grounded (DLG) technology, spoof surface plasmon polaritons (SSPPs) are used in the design of waveguide bandpass filters (BPFs). The SSPP unit cells' two-layer metallic strips are coated with a dielectric substrate and grounded to the wall of the rectangular waveguide (RWG), where the substrate is placed into the RWG's E-plane. The SSPP itself realizes the bandpass filtering responses as a result of the capacitance coupling between two layers that are introduced due to the DLG structure. The groove length can be changed flexibly to control the center frequency. Furthermore, an enhanced DLG-SSPP unit cell structure with a central patch is also designed to independently control the upper cut-off frequency [30]. Figure 3.8a shows the measurement setup and (b) the cavity used to design the BPF (c) DLG-SSPP (d) gapped DLG-SSPP (e) gapped DLG-SSPPs with the central patch.

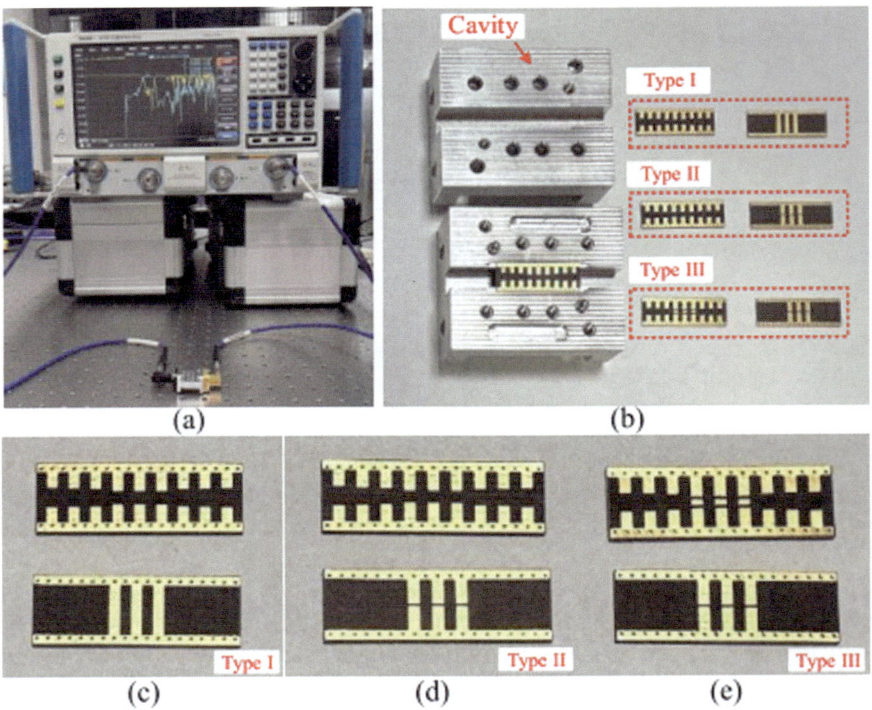

Fig. 3.8 **a** Measurement setup **b** cavity and proposed BPF **c** DLG-SSPP **d** gapped DLG-SSPP **e** gapped DLG-SSPPs with central patch [30]. Reproduced with permission from the IEEE

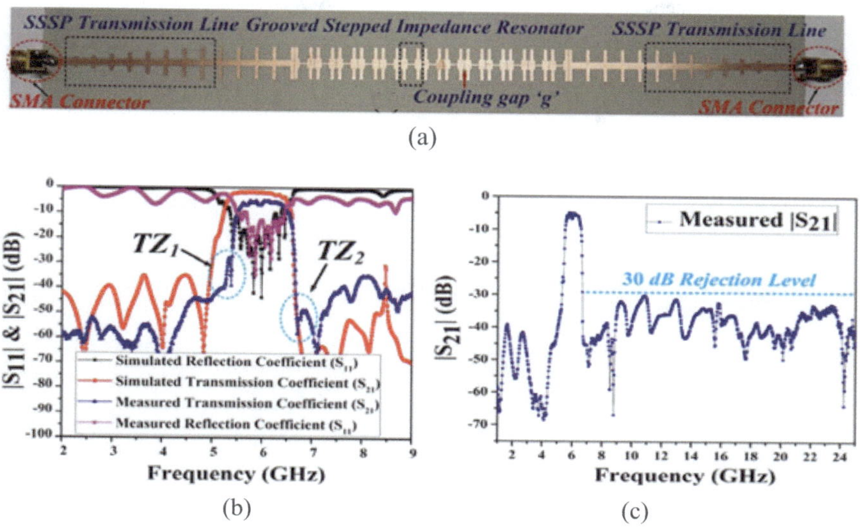

Fig. 3.9 **a** Fabricated coupled-gap GSIR BPF **b** measured and simulated S-parameters **c** spurious frequency response of BPF [31]. Reproduced with permission from the IEEE

Multi-mode grooved stepped impedance resonators (GSIRs) at the THz frequency regime enable an end-coupled broadband spoof plasmonic-based bandpass filter (BPF) at a subwavelength scale. The suggested GSIRs are externally coupled to the source and load via a spoof surface plasmon polaritons (SPP)-based transmission line, and they are end-coupled to one another [31]. The filter has a shape factor of 1.35, a 3-dB bandwidth of 0.35 THz (from 0.17 to 0.52 THz), and a fractional bandwidth (FBW) of 87.85%. Within the proposed filter's passband, the transmission and reflection coefficients are better than approximately 6 dB and 15 dB, respectively. Additionally, the designed filter's 60 dB spurious rejection bandwidth in the upper stopband is 1.3 THz (0.62–1.92 THz) as shown in Fig. 3.9.

In [32], the implementation of a W-band E-plane waveguide bandpass filters (BPFs) involves inserting dielectric substrates into standard rectangular waveguides to create millimeter-wave BPFs. Surface plasmon polariton (SSPP) patterns and grounded lines are coated on the front and back sides of inserted dielectric substrates. The dimensions of SSPP patterns and grounded lines can be adjusted flexibly and independently to control the upper and lower cut-off frequencies of millimeter-wave E-plane waveguide BPFs. Two distinct W-band E-plane BPFs using SSPPs and grounded lines are designed as shown in Fig. 3.10. Figure 3.11a, b shows the simulated and measured results of W-band E-plane waveguide BPF-A and BPF-B, respectively. Spoof surface plasmon polaritons (SSPPs) are the basis for a class of millimeter-wave E-plane waveguide bandpass filters (BPFs) that have been introduced [33]. To design BPFs, three different types of SSPPs with distinct patterns coated on the dielectric substrate are inserted into the WR-10 standard rectangular waveguide's E-plane. By varying the asymptotic frequencies of SSPP unit cells, the

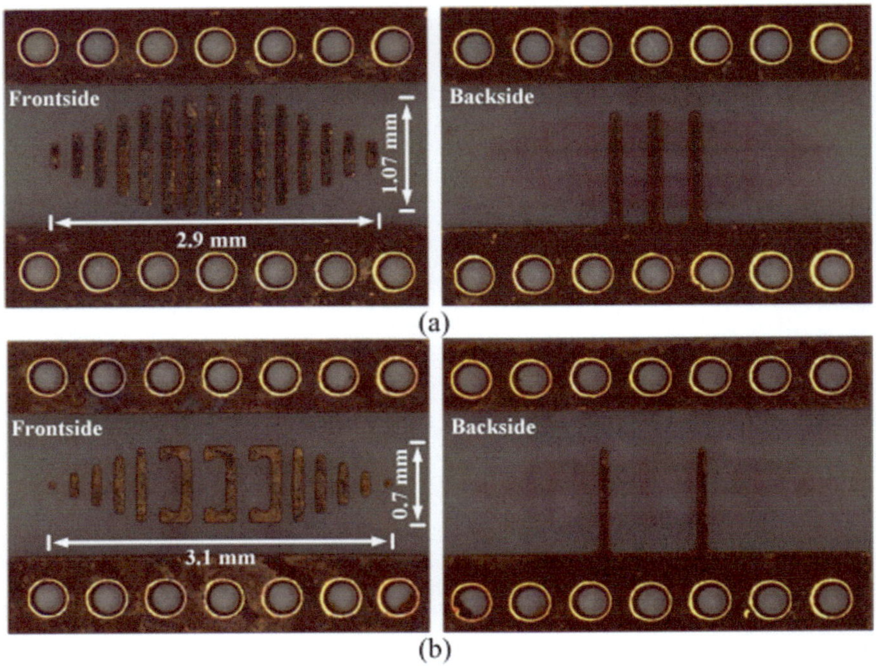

Fig. 3.10 **a**, **b** Photograph of fabricated W-band E-plane waveguide BPF-A and BPF-B, respectively [32]. Reproduced with permission from the IEEE

bandwidths and center frequencies of the suggested E-plane waveguide BPFs can be flexibly adjusted.

A dual-band bandpass filter with high Q has been designed using a corrugated ring resonator based on SSPP (BPF). Next, to provide loss compensation and amplification at the designated frequency bands of the filtering structure, a subwavelength scale amplifier has been embedded in the output port of the designed BPF [34].

A tri-band bandpass response is a result of the cut-off characteristic and multi-mode resonance of SSPPs. Every unit cell is a multi-mode resonator made up of two double-sided folded strips oriented in opposite directions. The ends of the folded strips are positioned near to each other to create coupled stubs. Under the cut-off frequency of the SSPP transmission structure, two distinct passbands are produced, separated by a sharp stopband. Because of the self-resonant mode of the suggested SSPPs unit cell, the resonant tunneling effect introduces the third passband above the cut-off frequency [35].

To realize a bandpass propagation, feature instead of a conventional low-pass one, spoof surface plasmon polaritons (SSPPs) are developed in a slotted half-mode substrate-integrated waveguide (HMSIW) by etching gradient slots on a tapered microstrip-to-HMSIW transition to create a compact transition. From 4.24 to 7.81 GHz, the measured 3-dB relative bandwidth is 59.25%. With a rejection level better than −31.6 dB, the upper rejection band measured can reach up to 16 GHz.

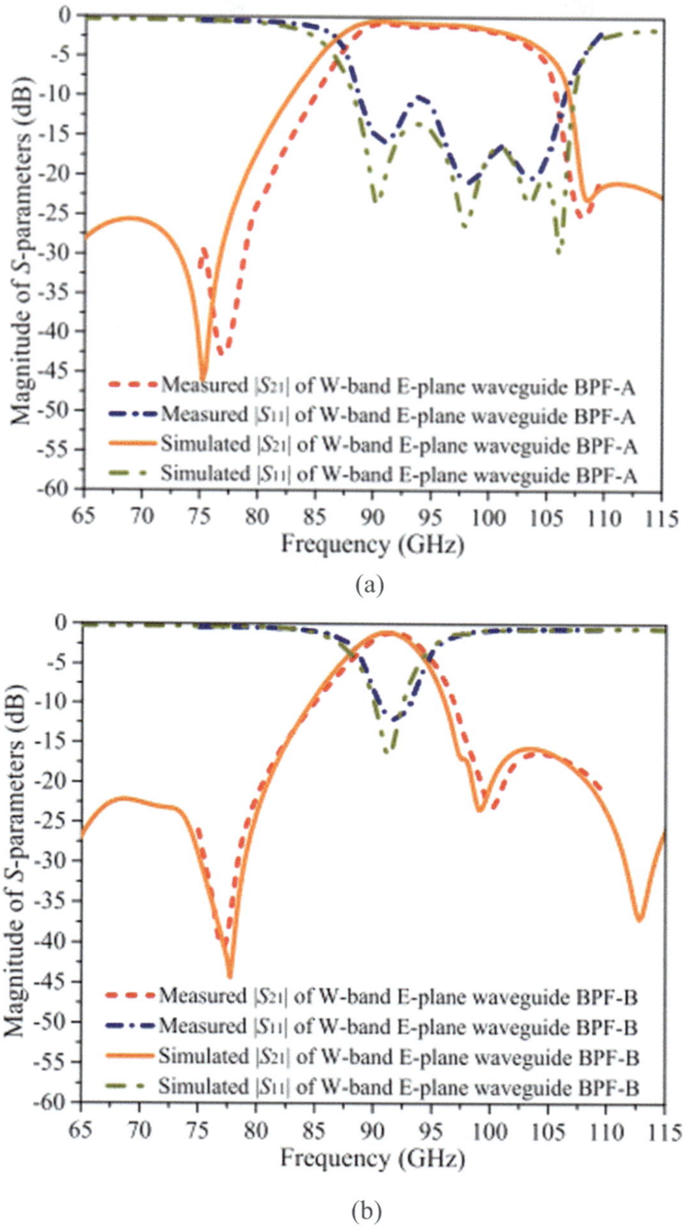

Fig. 3.11 **a**, **b** Simulated and measured results of W-band E-plane waveguide BPF-A and BPF-B, respectively [32]. Reproduced with permission from the IEEE

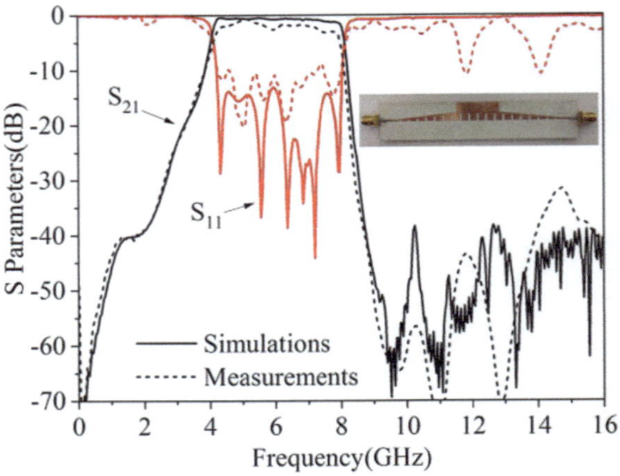

Fig. 3.12 Simulated and measured S-parameter with a photograph of BPF in the inset [36]. Reproduced with permission from the IEEE

The filter measures just $2.69\lambda g \times 0.42\lambda g$ in total [36]. Figure 3.12 shows the simulated and measured S-parameter of the bandpass filter with its fabricated prototype in the inset. A millimeter-wave broadband bandpass filter (BPF) based on slotted half-mode substrate-integrated waveguide (HMSIW) spoofs surface plasmon polaritons (SSPPs) using gallium arsenide (GaAs). Due to the SSPP unit cell's distinct dispersion properties, it is possible to precisely control the center frequency and bandwidth of this SSPP-based on-chip BPF by adjusting its geometric dimensions [37].

A substrate-integrated waveguide (SIW) and spoof surface plasmon polariton (SSPP) hybrid filter with broad out-of-band rejection and broadband bandpass capabilities is reported. The SSPP and SIW units' structural parameters can be altered independently to modify the passband's lower and upper cut-off frequencies, respectively. The passband range of a hybrid SSPP-SIW filter is 7.3–11.2 GHz, with a relative bandwidth of 44%. From 11.8 to 19.8 GHz, there is less than −40 dB of upper stopband rejection as shown in Fig. 3.12 [38]. A substrate-integrated plasmonic waveguide (SIPW) microwave bandpass filter consists of double arrays of slots etched on the top and bottom metal layers of a substrate-integrated waveguide (SIW) to support spoof surface plasmon polariton (SSPP) modes with low and high cut-off frequencies. This filter has a passband from 7.5 to 13.0 GHz with high return loss and low insertion loss shown in Fig. 3.13 [39]. Bandpass transmission can be achieved by etching periodic grooves at the top and bottom metal layers of a half-mode substrate-integrated waveguide (HMSIW) and double-layer spoof surface plasmon polaritons (SSPPs), which are composed of two corrugated metal strips. A parametric study is used to examine how significant parameters affect the suggested bandpass filter's performance. The bandpass filter attains a bandwidth of approximately 69.77% (15.6–32.1 GHz) for $|S_{11}|$ −0.8 dB [40].

The design of a dual-bandpass filter is based on a half-mode substrate-integrated waveguide (HMSIW) and spoof surface plasmon polaritons (SSPPs). To support SSPPs, this filter employs etched metal layers with periodic rectangular groove units. Concurrently, the combination of the substrate-integrated waveguide (SIW)

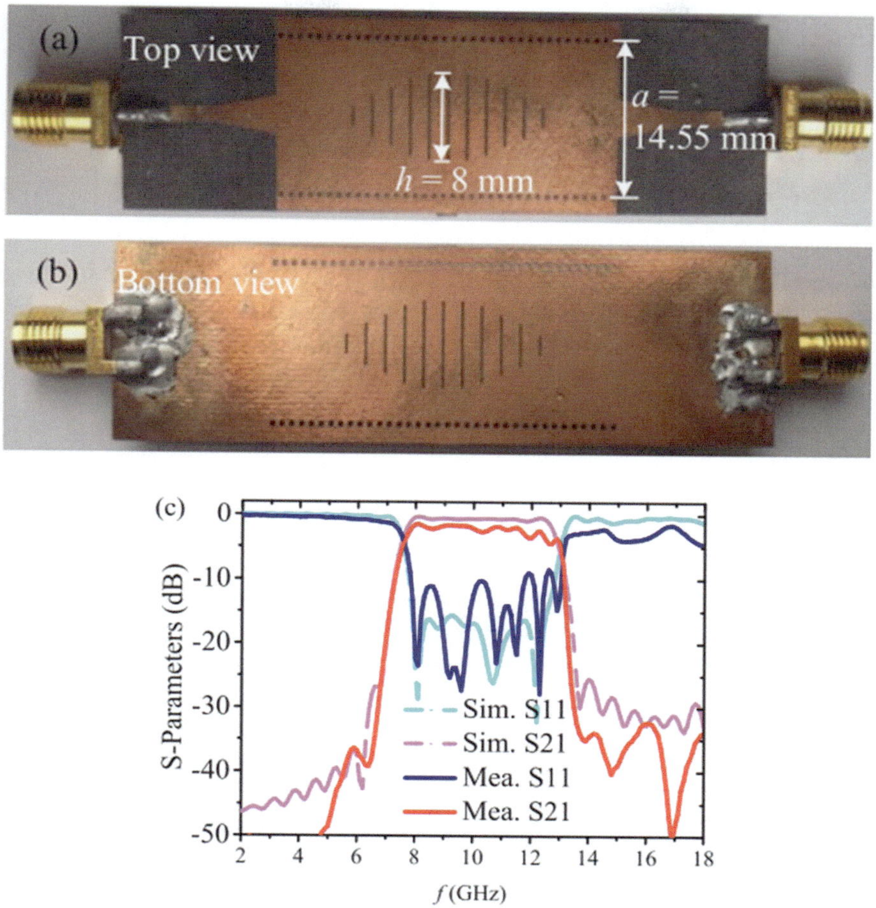

Fig. 3.13 Photograph of fabricated SIPW filter **a** top view and **b** bottom view. **c** Simulated and measured the S-parameter of the SIPW filter [39]

unit creates a strong coupling along the waveguide's propagation direction. Double rows of metal are added through holes in the gradient of the rectangular groove's transition section to restrict signal propagation and create the dual-bandpass filter's effect. The insertion loss of the dual-bandpass filter is less than 3 dB in the frequency from 8.7 to 16.4 GHz and 28.3 to 34.6 GHz, and the stopband rejection level is lower than −20 dB from 17.2 to 26.6 GHz [41]. Miniaturization and low insertion loss at the Ka-band are achieved by a narrowband bandpass filter made up of spoof surface plasmon polaritons (SSPPs) and substrate-integrated waveguide (SIW). Low and high cut-off responses are exhibited by SIWs and SSPPs, respectively, which can be combined to provide filtering functions. The cut-off frequencies of SIW and SSPPs can be tail-tailed to converge at a specific frequency for the narrowband filter, where the mode impedance ramps quickly and the SIW mode acts as a quasi-evanescent mode. By creating I-shaped SSPP structures with large effective length, a hybrid filter operating at 30 GHz with a 3-dB bandwidth as small as 3% and an insertion loss as low as 1.5 dB [42].

It has been possible to obtain UWB characteristics with fixed single and dual-notch bands. Usually, the band-notched feature was obtained by adding the proper filters to the UWB systems. These filters with fixed notch bands are susceptible to varying levels of in-band interference. Different transmission line technologies, including microstrip, strip line, substrate-integrated waveguide (SIW), SSPP, etc., were used to obtain the notches. The SSPP is a relatively new technology that is attracting a lot of interest because of its appealing characteristics, which include low nonlinear dispersion, low crosstalk, slow wave, and subwavelength field confinement. To improve the overall characteristics of the system, passive SSPP-based components, like filters, can be implemented and integrated with another SSPP structure or transmission line.

3.4 Techniques to Design Band-Notched Filter

In addition to ultra-wideband applications [43], on-chip-integrated terahertz circuits use band-notch filters. Under the fundamental mode and high-order mode, the grooves corrugated metallic SSPPs transmission line (TL) with etched complementary split ring resonators (CSRRs) exhibits a pass-reject-pass characteristic in the frequency domain. SRRs with various resonance frequencies are loaded into the SSPPs TL grooves to increase the rejection band and isolation of the filter. The filter's isolation is greater than 10 dB and can reach a maximum isolation of 50 dB within the frequency range of 128–158 GHz with a maximum relative bandwidth of up to 21% [44].

Surface plasmonic polaritons (SSPPs) can be efficiently and strongly constrained in microwave transmission thanks to a folded line spoof waveguide. The folded line SSPP waveguide's lateral dimension can be cut by 56% when compared to a traditional rectangular-grooved SSPP waveguide with the same cut-off frequency as shown in Fig. 3.14. Better miniaturization benefits and stronger field confinement are found in the folded line SSPP waveguide. The high-order mode of the suggested unit serves as the basis for the design of a dual-band bandpass filter. An asymmetric structure is then added to produce a tri-band bandpass filter with good performance. Furthermore, by adding an interdigital finger to the folded line unit, a bandpass filter with a wide stopband between two passbands is created, increasing the stopband's width by 50% in comparison to the filter without an interdigital finger. Two notch bands and two passbands of a dual-band passband filter are obtained, respectively, by optimizing the length of the interdigital finger [45].

To create a bandpass filter (BPF) with a notched band, the higher-order mode of spoof surface plasmon polaritons (SSPPs) is taken into consideration. The first higher-order mode (Mode 1) of SSPPs is the basis for the development of the high-efficiency passband. Since Mode 1 has an inherent bandpass response while the fundamental mode has a low-pass response, the BPF it generates can be excited

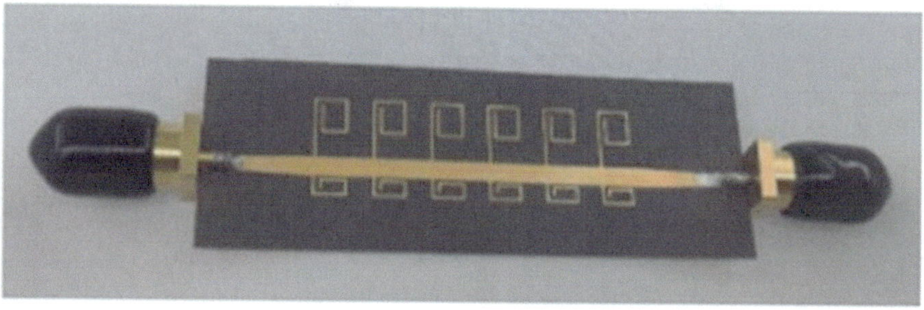

(a)

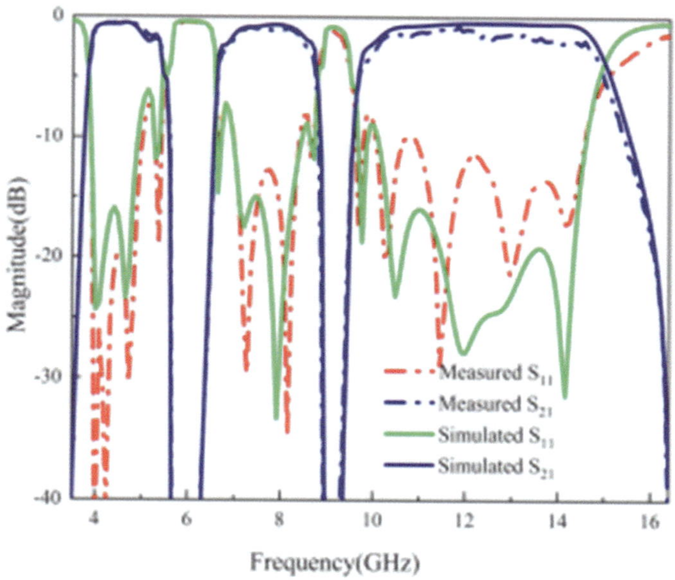

(b)

Fig. 3.14 **a** Photograph of dual-band-notched filter **b** simulated and measured S-parameters [45]

without the need for an additional high-pass filtering structure or a gradient transition structure. A BPF with a notched band is created by integrating a U-shaped resonator into the SSPP structure. As the resonator's length increases, the notched band's location will redshift [46].

Comparing the plasmonic waveguides to traditional rectangular-grooved plasmonic waveguides reveals much tighter terahertz field confinement and much lower asymptotic frequencies. A structure with the same asymptotic frequency but only 25% of the lateral dimensions of conventional spoof surface plasmon polariton waveguides can be designed. Ultimately, we designed an ultra-wideband low-pass filter with low insertion loss and sharp roll-off (cut-off frequency at 1.29 THz) by combining this architecture with broadband couplers. Additionally, a planar plasmonic notch filter with rejection of more than 17 dB between 0.97 and 0.99 THz is demonstrated by introducing double-ring resonators based on spiral-shaped units [47]. Using double-sided corrugated metallic strips, the vertical split ring resonator (VSRR) is added to the planar spoof surface plasmon polariton (SPP) waveguide to create the plasmonic filter. The SPP propagation is inhibited near the z-component of the electric field of the spoof SPP mode, which excites the second resonance of the VSRR [48].

Split ring resonators (SRRs) are loaded on the transmission line of spoof surface plasmon polaritons (SPPs) to create an ultra-compact rejection filter. We demonstrate the mechanism of the rejection characteristic for this compact filter with clear results from the dispersion analysis of the spoof SPP transmission line both with and without the SRR loading. At rejection frequencies, the isolation of both filters can be as low as −40 dB, and in some cases, less than −20 dB [49]. One important ecosystem of SSPPs to be designed is band rejection filters, particularly multi-band rejection filters. Through the insertion of interdigital capacitance loaded loop resonators (IDCLLRs) into the grooves of the corrugated SSPP transmission line (TL), compact multi-band rejection filters, including dual-band and triple-band, can be created. When measured against conventional metamaterial structures of the same size, such as split ring resonators (SRRs) and complementary SRRs (CSRRs), IDCLLRs exhibit superior qualities such as a larger dynamic range, more tunable parameters, and a lower resonant frequency. These attributes allow the device to be more compact and provide greater flexibility in terms of bandwidth adjustment. With a maximum relative bandwidth of 9.5% for dual-band filters and 8.8% for triple-band filters, it can offer a broad stop-bandwidth. and high isolation (both filters' isolation can be less than −30 dB, with the dual-band filter's maximum isolation reaching −60 dB and the triple-band filter's maximum isolation reaching −58 dB) [50].

An SSPP-based transmission line and six-coupled circular ring resonators (CCRRs) etched among the ground planes of the center corrugated strip comprise a novel single-layer dual-band rejection filter based on SSPPs. The SSPP structure's electric field excites these resonators. The strip's added ground on both sides produces electromagnetic fields that are more tightly spaced, which enhances the filter's lower frequency performance. The overall filter size is reduced significantly and the mode conversion efficiency at the transition from coplanar waveguide (CPW) to the SSPP line is increased by eliminating flaring ground in comparison to common SSPP-based constructions. The X-and K-bands contain two prohibited bands: 8.6–11.2 GHz and 20–21.8 GHz [51]. Figure 3.15a shows the fabricated prototype of band-notched filter whereas (b) shows the simulated and measured S-parameters of the band-notched filter.

The dielectric substrate, metal ground plane, and U-shaped SSPP transmission line embedded in 山-shaped resonator structures make up the narrow wideband (NWB) BPF's design. The transmission zero was produced in a wideband range using the loaded-shaped resonator structure. The designed NWB BPF has a frequency range

(a)

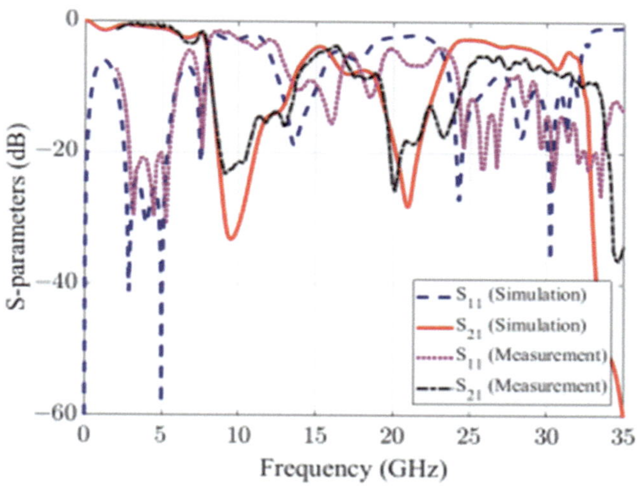

(b)

Fig. 3.15 **a** Fabricated prototype of band-notched filter **b** simulated and measured S-parameters [51]

of 4.35–8.12 GHz for operation and a notched band of 6.2–6.7 GHz with a rejection of over −20 dB. Both the passband cut-off frequency and the notched band position can be independently changed by modifying the geometric parameters of the NWB BPF [52].

A band-stop filter with etching gradient semi-circular grooves on both sides of the coplanar waveguide's middle strip line is based on spoof surface plasmon polaritons for high-efficiency band rejections. By varying the unit cell's length and the semi-circular holes' radii, one can regulate the band-stop frequency and operating bandwidth, respectively. In order to analyze how important parameters affect the band-stop filter's performance and make practical designs, parametric studies are carried out [53]. A notch filter for an ultra-wideband microwave waveguide based on spoof surface plasmon polaritons is proposed. Such a filter has an advantage in structure size without performance loss because of subwavelength confinement. In order to suppress both WLAN and satellite communication interference at the same time, a spoof SPP-based notch is introduced. The cut-off frequency can reach 20 GHz, and both the notch frequency and the cut-off frequency are dependent on the structure parameters. To efficiently couple energy into spoof SPP waveguide, an adiabatic transition based on gradient hole-size and flaring ground is designed. To avoid the current WLAN and satellite communication signals, the result reveals a cut-off frequency of 17.4 GHz with an insertion loss better than 3 dB throughout the entire passband and more than 20 dB rejections at 5.36 GHz and 9.32 GHz with 10 dB fractional bandwidth 1.07% and 0.74%, respectively [54].

Using the interactions between spoof SPP waveguide and metamaterial structures, multiple band rejection filters based on SPPs are designed in dual frequency bands. The spoof SPP waveguide is decorated with π-type etchings to achieve the multi-band rejections. More importantly, by adjusting various geometrical parameters of the π-type decorations, the bandwidths, isolation, and shifting of the rejection frequencies of filters can be controlled very effectively [55]. For ultra-wideband (UWB) applications, a controllable band-stop filter based on spoof surface plasmon polaritons (SSPPs) using complementary split ring resonators (CSRRs) is used. For the ground plane to have the necessary band rejection characteristics, four CSRRs are inserted. The depth of the SSPP grooves can be adjusted to control the filter cut-off frequency. The geometric parameters and position of the CSRRs are used to control the stopband's frequency and bandwidth, respectively. Adjusting the stopband has no effect on the cut-off frequency because the stopband control process operates independently of the filter cut-off frequency. The 20 dB rejection bandwidth with the CSRR diameters of 3.2 and 4 mm are 10.5–11.2 GHz and 7.8–8.6 GHz, respectively, according to two filter prototypes with different CSRR diameters. The cut-off frequency is about 13.1 GHz for both filters, which indicates that stopband control does not change the cut-off frequency [56]. Cascading spoof surface plasmon polaritons (SSPPs) with different rejection bands but the same structure results in the implementation of a tri-band rejection filter and a dual-band rejection filter.

The filters offer better independent tunability, wider rejection bands, and more compact structures than traditional ones. It is possible to independently adjust the filter's cut-off frequency, rejection bandwidth, and rejection band center frequency. Additionally, there is independent control over the various rejection bands contained within the same filter. There are also presented agreements between the S_{21} of filters and the dispersion of SSPP units. The measurement results show that all of the rejection bands are located in the Ku- and K-bands, and that both filters load multiple rejection bands on the wide low-passband of 27.7 GHz. The average rejection bandwidth and the average rejection depth of the two filters are 1.49 GHz and 42.1 dB, respectively [57].

A technique utilizing metamaterial structures to regulate the spoof SPP rejection rates is reported in the literature. The SPP modes are rejected close to the resonant frequencies within the propagating band by introducing electrically resonant metamaterials near an ultrathin corrugated metallic strip, or the spoof SPP waveguide, to create tight coupling and mismatch of surface impedance. We control the rejections of SPP modes from narrowband to broadband by varying the scaling factor of metamaterial structures [58]. At its resonant frequency, a band-stop filter featuring single-loop split ring resonators (SRRs) can achieve a miniaturized size of $0.052\lambda_0 \times 0.278\lambda_0$. The rectangular corrugations in the traditional spoof surface plasmon polariton provide a low-pass response, and the resonance of the SRRs induces an additional band-stop response. With a maximum attenuation of 31 dB and a high Q-factor of 93, the stop band is centered at 49 GHz and has a -10 dB bandwidth of 4.1 GHz. In the passband, the filter has a low insertion loss of less than 2.8 dB [59].

A novel type of capacitive-coupled series spoof SPPs introduces the stopband. A new unit cell is created from two conventional H-shaped unit cells, and each time a new unit cell is created, a gap of a specific length separates them in order to implement capacitive coupling. During the transmission of SPPs, the capacitive coupling throws off the original surface impedance matching, resulting in the stopband. The structure maintains the bandpass feature in 5–9 GHz and 9.5–11 GHz but exhibits a stopband in 9–9.5 GHz. Figure 3.16a shows the fabricated prototype of band-notched filter, whereas Fig. 3.16b shows the reflection and transmission coefficient of band-notched filter [60].

The transmission loss is approximately 3 dB and the reflection coefficient is less than -10 dB in the passband; in the stopband, the transmission coefficient is less than -30 dB and the reflection coefficient is -2 dB [60]. Capacitively coupled unit cells are incorporated into the SSPP line to introduce a notch band [61]. Figure 3.17 shows the geometry of the proposed SSPP transmission line; in this figure, the zoomed part is the incorporated capacitively coupled unit cell where 'g' is the gap between the capacitively coupled unit cell and transmission line while 'h' is the height of the unit cell. The capacitively coupled unit cell is placed after every three unit cells (at a gap of 12 mm).

The S-parameters of the SSPP line with capacitively coupled unit cells are shown in Fig. 3.18. The S-parameters of the line show a rejection band in the frequency range of 8.7–9.9 GHz. The impact of varying the gap (g) is also shown in Fig. 3.18. Here, there is a compromise between the notch depth and insertion loss after the notch band. When $g = 1$ mm, the notch depth is better but the insertion loss is poorer just below the cut-off frequency of the line. The optimized value of gap is taken as $g = 0.1$ mm.

The SSPP transmission line (with a notch band) of a size 227 mm $\times$ 45 mm is fabricated on a 1.5 mm thick F4B substrate with $\varepsilon_r = 2.65$ and $\tan\delta = 0.002$ to validate the performance. The reflection and transmission coefficients of the SSPP transmission line are measured using a vector network analyzer (PNA-E8364C). Figure 3.19 plots the measured reflection and transmission coefficients. The reflection coefficient is below -10 dB in the passband region and greater than -1 dB in the stop band region. Similarly, the transmission coefficient is below -20 dB in the stop band region, while in the passband region, it is below -1 dB at lower frequencies and -2.6 dB at higher frequencies near the cut-off region of the SSPP line. A small discrepancy in simulated and measured results is due to the fabrication imperfection.

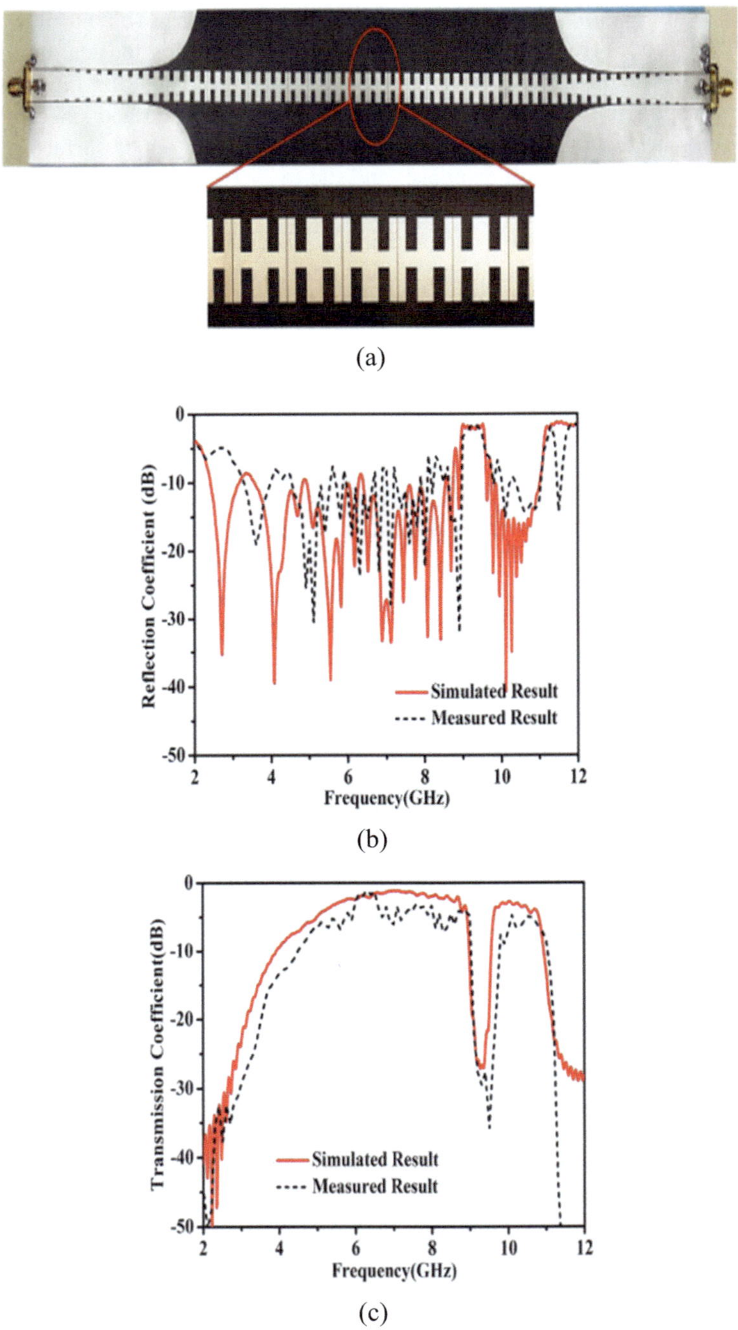

Fig. 3.16 **a** Fabricated prototype of the band-notched filter **b** reflection and **c** transmission coefficient of band-notched filter [60]

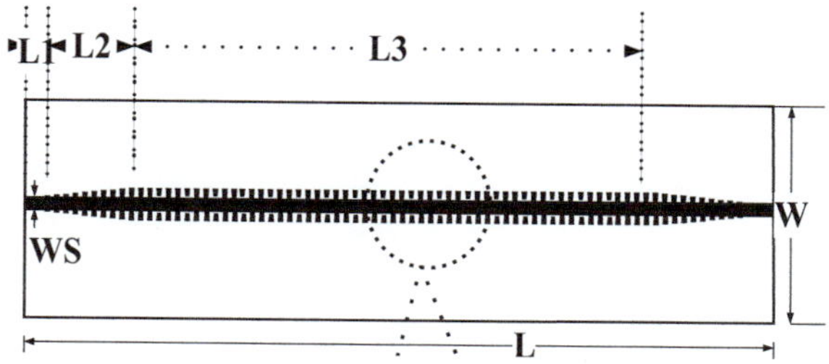

Fig. 3.17 SSPP transmission line with capacitively coupled unit cells incorporated in between the line to create a notch band

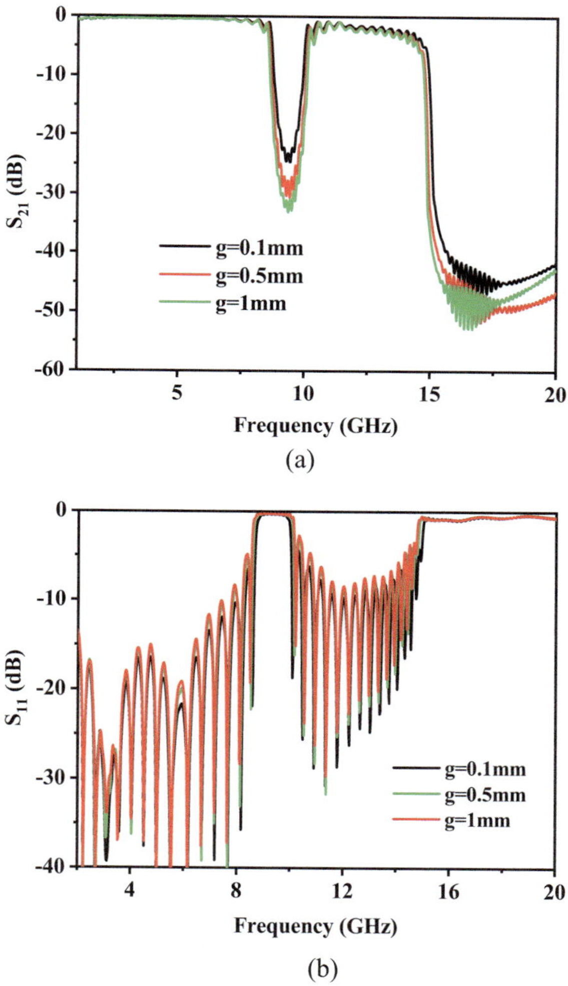

Fig. 3.18 S-parameters of the SSPP line after incorporating capacitively coupled unit cells, also showing the impact of varying 'g' on **a** S_{21} **b** S_{11}

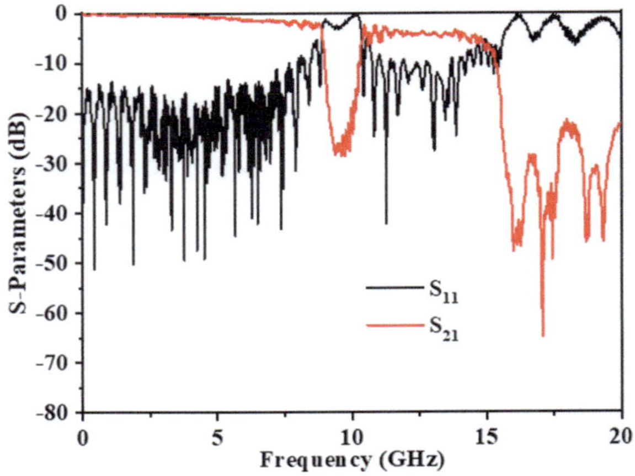

Fig. 3.19 Measured S-parameters of the SSPP line with capacitively coupled unit cells at $g =$ 0.1 mm

3.5 Conclusion

The filter design using spoof surface plasmon polariton and their advantages such as field confinement and subwavelength resolution are discussed in this chapter. Bandpass filter and band-notched filter applications in microwave and terahertz domains are presented. Different techniques used to design bandpass and band-notched filters using spoof surface plasmon polariton are explained in detail. Multi-band design of band-notched and bandpass filter is also presented.

References

1. Ma, H.F., Shen, X., Cheng, Q., Jiang, W.X., Cui, T.J.: Broadband and high-efficiency conversion from guided waves to spoof surface plasmon polaritons. Laser Photon. Rev. **8**(1), 146–151 (2014)
2. Zhang, D., Zhang, K., Wu, Q., Ding, X., Sha, X.: High-efficiency surface plasmonic polariton waveguides with enhanced low-frequency performance in microwave frequencies. Opt. Express **25**(3), 2121 (2017)
3. Sharma, S., Singh, R.K., Basu, A., Koul, S.K.: Spoof surface plasmon polariton based frequency scanning antennas. IETE J. Res. (2023). https://doi.org/10.1080/03772063.2023.2252399
4. Sharma, S., Singh, R.K., Basu, A., Koul, S.K.: A wide-band transition from microstrip line to microstrip spoof surface plasmon polariton line for microwave/millimeter wave applications. Int. J. RF Microw. Comput. Aided Eng. (RFCAD) (2022)
5. Sharma, S., Singh, R.K., Basu, A., Koul, S.K.: A wide-angle forward to backward scanning wideband leaky-wave antenna based on microstrip spoof surface plasmon polaritons. IET Microw. Antennas Propag. 1–12 (2021). https://doi.org/10.1049/mia2.12213
6. Zhang, D., Zhang, K., Wu, Q., Yang, G., Sha, X.: High-efficiency broadband excitation and propagation of second-mode spoof surface plasmon polaritons by a complementary structure. Opt. Lett. **42**(14), 2766–2769 (2017)
7. Zhang, D., Zhang, K., Wu, Q., Dai, R., Sha, X.: Broadband high order mode of spoof surface plasmon polaritons supported by compact complementary structure with high efficiency. Opt. Lett. **43**(13), 3176–3179 (2018)

8. Gao, X., Che, W., Feng, W.: Novel non-periodic spoof surface plasmon polaritons with H-shaped cells and its application to high selectivity wideband bandpass filter. Sci. Rep. **8**, 2456 (2018)
9. Zhang, Q., Zhang, H.C., Wu, H., Cui, T.J.: A hybrid circuit for spoof surface plasmons and spatial waveguide modes to reach controllable band-pass filters. Sci. Rep. **5**, Art. no. 16531 (2015)
10. Lin, Z., Li, Y., Li, L., Zhao, Y.-T., Xu, J., Chen, J.: Miniaturized bandpass filter based on high-order mode of spoof surface plasmon polaritons loaded with capacitor. IEEE Trans. Plasma Sci. **51**(1), 254–260 (2023). https://doi.org/10.1109/TPS.2022.3232850
11. Liu, Y., Xu, K.-D., Guo, Y.-J., Chen, Q.: High-order mode application of spoof surface plasmon polaritons in bandpass filter design. IEEE Photonics Technol. Lett. **33**(7), 362–365 (2021)
12. Liu, L., Hu, Y., Li, R.-Z., Zhang, Q., Zhu, L.: Compact and narrow-band bandpass filter using spoof surface plasmon polaritons. IEEE Photonics Technol. Lett. **33**(13), 676–679 (2021)
13. Wang, R., et al.: J. Phys. D Appl. Phys. **52**, 325101 (2019)
14. Zhu, S., Wen, P., Liu, Y.: A compact filter based on spoof surface plasmon polariton waveguide for wide stopband suppression. IEEE Photonics Technol. Lett. **34**(9), 475–478 (2022)
15. Mittal, G., Pathak, N.P.: Spoof surface plasmon polaritons based microwave bandpass filter. Microw. Opt. Technol. Lett. **63**, 51–57 (2021)
16. Wang, M., Wei, G., Han, K.: Single-conductor spoof surface plasmon polaritons transmission line with bandpass feature based on embedded cross-shaped resonators. IEEE Trans. Circuits Syst. II Express Briefs **70**(12), 4359–4363 (2023)
17. Liu, L., Liu, Y., Jin, X., Li, Z., Tian, Y.: A broad dual-band bandpass filter design based on double-layered spoof surface plasmon polaritons. Electronics **11**, 697 (2022)
18. Wei, Y., Wu, Y., Wang, W., Pan, L., Yang, Y., Liu, Y.: Double-sided spoof surface plasmon polaritons-line bandpass filter with excellent dual-band filtering and wide upper band suppressions. IEEE Trans. Plasma Sci. **48**(12), 4134–4143 (2020)
19. Zhang, X., Liu, S., Wang, L., et al.: Design and analysis of ultra-wideband and miniaturized bandpass filter based on spoof surface plasmon polaritons. Plasmonics **17**, 789–797 (2022)
20. Zhao, L., Zhang, X., Wang, J., et al.: A novel broadband band-pass filter based on spoof surface plasmon polaritons. Sci. Rep. **6**, 36069 (2016)
21. Jidi, L., Cao, X., Gao, J., Yang, H., Li, S., Li, T.: An ultra-thin and compact band-pass filter based on spoof surface plasmon polaritons. IEEE Access **8**, 171416–171422 (2020)
22. Feng, W., Shi, S., Yin, S., Zhu, H., Shi, Y., Che, W.: Novel wideband bandpass filters using double-sided quasi-SSPPs transmission line. IEEE Trans. Circuits Syst. II Express Briefs **69**(7), 3174–3178 (2022). https://doi.org/10.1109/TCSII.2022.3169550
23. Chen, Z.-M., et al.: A high efficiency band-pass filter based on CPW and quasi-spoof surface plasmon polaritons. IEEE Access **8**, 4311–4317 (2020)
24. Yan, S., et al.: A terahertz band-pass filter based on coplanar-waveguide and spoof surface plasmon polaritons. IEEE Photonics Technol. Lett. **34**(7), 375–378 (2022)
25. Pan, L., Wu, Y., Wang, W., Wei, Y., Yang, Y.: A flexible high-selectivity single-layer coplanar waveguide bandpass filter using interdigital spoof surface plasmon polaritons of bow-tie cells. IEEE Trans. Plasma Sci. **48**(10), 3582–3588 (2020)
26. Feng, W., Feng, Y., Shi, Y., Shi, S., Che, W.: Novel differential bandpass filter using spoof surface plasmon polaritons. IEEE Trans. Plasma Sci. **48**(6), 2083–2088 (2020)
27. Li, Q., Guo, D., Li, J., Xu, K.-D.: W-band bandpass filter using spoof surface plasmon polaritons and double-grating rectangular waveguide. IEEE Microw. Wirel. Technol. Lett. **33**(10), 1426–1429 (2023)
28. Li, Q., Guo, D., Mou, J., Li, J., Xu, K.-D.: Groove gap waveguide bandpass filter based on spoof surface plasmon polariton for Ka-band applications. IEEE Trans. Microw. Theory Tech. **72**(1), 340–347 (2024)
29. Guo, Y.J., Xu, K.D., Liu, Y., Tang, X.: Novel surface plasmon polariton waveguides with enhanced field confinement for microwave-frequency ultra-wideband bandpass filters. IEEE Access **6**, 10249–10256 (2018)

30. Feng, Y., Xu, K.-D., Niu, Z., Zhang, B., Liu, L., Fan, Y.: Ka-band waveguide bandpass filters using double-layer grounded spoof surface plasmon polaritons. IEEE Trans. Microw. Theory Tech. https://doi.org/10.1109/TMTT.2024.3373504
31. Jaiswal, R.K., Pandit, N., Pathak, N.P.: Spoof plasmonic-based band-pass filter with high selectivity and wide rejection bandwidth. IEEE Photonics Technol. Lett. **31**(15), 1293–1296 (2019)
32. Liu, Y., Xu, K.-D.: W-band E-plane waveguide bandpass filters using spoof surface plasmon polaritons and grounded lines. IEEE Trans. Microw. Theory Tech. **72**(6), 3564–3576 (2024)
33. Liu, Y., Xu, K.-D., Li, J., Guo, Y.-J., Zhang, A., Chen, Q.: Millimeter-wave E-plane waveguide bandpass filters based on spoof surface plasmon polaritons. IEEE Trans. Microw. Theory Tech. **70**(10), 4399–4409 (2022)
34. Jaiswal, R.K., Pandit, N., Pathak, N.P.: Amplification of propagating spoof surface plasmon polaritons in ring resonator-based filtering structure. IEEE Trans. Plasma Sci. **48**(9), 3253–3260 (2020)
35. Zhao, H., et al.: Tri-band band-pass filter based on multi-mode spoof surface plasmon polaritons. IEEE Access **8**, 14767–14776 (2020)
36. Zhang, D., Zhang, K., Wu, Q., Jiang, T.: A compact wideband filter based on spoof surface plasmon polaritons with a wide upper rejection band. IEEE Photonics Technol. Lett. **32**(24), 1511–1514 (2020)
37. Guo, Y.-J., Xu, K.-D., Deng, X., Cheng, X., Chen, Q.: Millimeter-wave on-chip bandpass filter based on spoof surface plasmon polaritons. IEEE Electron. Device Lett. **41**(8), 1165–1168 (2020)
38. Chen, P., Li, L., Yang, K., Chen, Q.: Hybrid spoof surface plasmon polariton and substrate integrated waveguide broadband bandpass filter with wide out-of-band rejection. IEEE Microw. Wirel. Compon. Lett. **28**(11), 984–986 (2018)
39. Ye, L., Chen, Y., Xu, K.D., Li, W., Liu, Q.H., Zhang, Y.: Substrate integrated plasmonic waveguide for microwave bandpass filter applications. IEEE Access **7**, 75957–75964 (2019)
40. Zhao, L., Li, Y., Chen, Z.M., et al.: A band-pass filter based on half-mode substrate integrated waveguide and spoof surface plasmon polaritons. Sci. Rep. **9**, 13429 (2019)
41. Lin, Y., Xiao, B., Li, X.: Dual bandpass filter based on spoof surface plasmon polaritons and half-mode substrate integrated waveguide. Opt. Quant. Electron. **54**, 12 (2022). https://doi.org/10.1007/s11082-021-03380-2
42. Zhang, B., et al.: Narrowband SIW-SSPP hybrid bandpass filter with compact profile at Ka-band. IEEE Access **11**, 98305–98314 (2023)
43. Revision of Part 15 of the Commission's Rules Regarding Ultra-Wideband Transmission System. FCC, Washington, DC, ET-Docket 98-153, First Note and Order (2002)
44. Zhu, H., et al.: Compact terahertz on-chip filter with broadband rejection based on spoof surface plasmon polaritons. IEEE Electron Device Lett. **43**(6), 970–973 (2022)
45. Yang, S., et al.: Compact muti-frequency band-pass filters based on high-order mode of folded line spoof surface plasmon polaritons waveguide. IEEE Access **12**, 7244–7254 (2024)
46. Xu, K.-D., Lu, S., Guo, Y.-J., Chen, Q.: High-order mode of spoof surface plasmon polaritons and its application in bandpass filters. IEEE Trans. Plasma Sci. **49**(1), 269–275 (2021)
47. Ye, L., et al.: Super subwavelength guiding and rejecting of terahertz spoof SPPs enabled by planar plasmonic waveguides and notch filters based on spiral-shaped units. J. Lightw. Technol. **36**(20), 4988–4994 (2018)
48. Yang, Z., Zhang, B., Chen, W., Yang, T.: Rejection of spoof SPPs using the second resonant mode of vertical split-ring resonator. IEEE Microw. Wirel. Compon. Lett. **29**(1), 23–25 (2019)
49. Zhao, S., Zhang, H.C., Zhao, J., et al.: An ultra-compact rejection filter based on spoof surface plasmon polaritons. Sci. Rep. **7**, 10576 (2017)
50. Wang, Z.X., et al.: J. Phys. D: Appl. Phys. **52**, 025107 (2019)
51. Farokhipour, E., Mehrabi, M., Komjani, N., Ding, C.: A spoof surface plasmon polaritons (SSPPs) based dual-band-rejection filter with wide rejection bandwidth. Sensors **20**, 7311 (2020)

52. Sun, S., Cheng, Y., Luo, H., et al.: Notched-wideband bandpass filter based on spoof surface plasmon polaritons loaded with resonator structure. Plasmonics **18**, 165–174 (2023)
53. Zhao, L., Liu, S., Wang, J., Shen, X., Cui, T.J.: Band-stop filter based on spoof surface plasmon polaritons. Electron. Lett. **55**, 607–609 (2019)
54. Xiao, B., Kong, S., Xiao, S.: Spoof surface plasmon polaritons based notch filter for ultra-wideband microwave waveguide. Opt. Commun. **374**, 13–17. ISSN 0030-4018
55. Aziz, A., et al.: J. Opt. **22**, 015001 (2020)
56. Homayoon, F., Heidari, A.-A.: A band-stop filter based on spoof surface plasmon polaritons using complementary split-ring resonators. Int. J. RF Microw. Comput. Aided Eng. **32**(7), 23186 (2022)
57. Li, L., Dong, L., Chen, P., Yang, K.: Multi-band rejection filters based on spoof surface plasmon polaritons and folded split-ring resonators. Int. J. Microw. Wirel. Technol. **11**(8), 774–781 (2019)
58. Pan, B.C., Liao, Z., Zhao, J., Cui, T.J.: Controlling rejections of spoof surface plasmon polaritons using metamaterial particles. Opt. Express **22**, 13940–13950 (2014)
59. Ling, H., Zhang, Y., Qian, P., et al.: Spoof surface plasmon polariton band-stop filter with single-loop split ring resonators. Int. J. RF Microw. Comput. Aided Eng. (2020)
60. Yin, J., Ren, J., Zhang, H., et al.: Capacitive-coupled series spoof surface plasmon polaritons. Sci. Rep. **6**, 24605 (2016)
61. Sharma, S., Singh, R.K., Basu, A., Koul, S.K.: Capacitively coupled spoof surface plasmon unit cells based transmission line with band rejection capability. In: 2022 IEEE Microwaves, Antennas, and Propagation Conference (MAPCON), Bangalore, India, pp. 1861–1864 (2022)

Chapter 4
Tunable Filter Using Spoof Surface Plasmon Polariton

4.1 Introduction

Filter plays a vital role in the design of an efficient communication system as already discussed in the previous chapter. In the previous chapter, we discussed about the bandpass and band-notched characteristics of the filter. The discussion was limited to the fixed bandpass or band-notched characteristic. But in the present wireless technology, the need for reconfigurability has increased so that with a single circuit, we can work on a wide range of applications. Ultra-wideband systems are one the most important systems that find application in IoT devices, healthcare monitoring, and on-body devices. Many wireless technologies coexist in this range such as WiMAX, WLAN, C-band, X-band, ITU band, and so on. Due to the overlapping of other wireless communication systems' spectrum with UWB systems, interference increases. The UWB with reconfigurable characteristics in UWB systems plays an important role in suppressing the interfering signals and helps in reducing the need for multiple antennas for diverse applications. Reconfigurable circuits with microstrip technology are already reported in the literature. The advantages of spoof surface plasmon polariton (SSPP) over microstrip lines such as field confinement, low crosstalk, and wavelength resolution characteristics make them perfect candidate for the present and future applications. They are easy to fabricate, have planar structure, and offer lower fabrication costs like that of a microstrip line. The highly sensitive characteristics of the SSPP help to achieve a wide tuning range and can be easily integrated with the active circuits. So, reconfigurable circuits with a wide tuning range can be designed with the help of SSPP technology. In this chapter, we will discuss different techniques used to design tunable bandpass and band-notched filters.

S. K. Koul and S. Sharma, *Microwave Integrated Circuits Based on Spoof Surface Plasmon Polariton*, Signals and Communication Technology,
https://doi.org/10.1007/978-981-96-9041-1_4

4.2 Tunable Bandpass Filter

The electronic tuning of bandpass filters can be obtained with the use of a p-i-n, varactor, or Schottky diode. Tunable graphene also plays a vital role in the design of tunable circuits. The metamaterial structures with SSPP technology are widely used to design tunable structures. The use of all these techniques to design a tunable bandpass filter will be discussed in this section.

The spoof surface plasmon polaritons (SSPPs) and coplanar waveguide (CPW)-based coupling structure is used to design a mechanically controllable bandpass filter [1]. Figure 4.1 shows the photograph of the fabricated prototype. The SSPP structure and CPW-based coupling structure are used to control the high cut-off frequency and filtering the low-frequency wave, respectively, which means that the low and high cut-off frequencies of the bandpass filter can be independently controlled. Further to improve the high-frequency performance, the split ring resonators are added to the structure. Figure 4.2 shows that the bandpass filter achieves a bandwidth (for $|S_{11}| < -10$ dB and $|S_{21}| > -1.6$ dB) of 35.9% (7.31–10.51 GHz).

The circuit discussed in the previous paragraph does not solve the problem of reconfigurability as it was mechanically controllable. So, a spoof surface plasmon

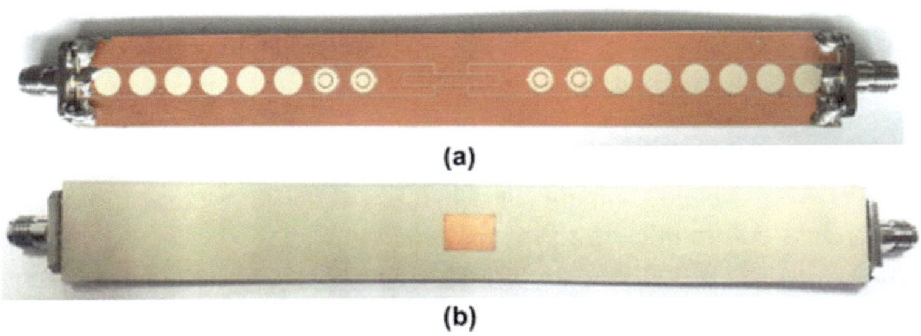

Fig. 4.1 Photograph of fabricated bandpass filter **a** top view **b** bottom view [1]

Fig. 4.2 Simulated and measured S-parameter of the bandpass filter [1]

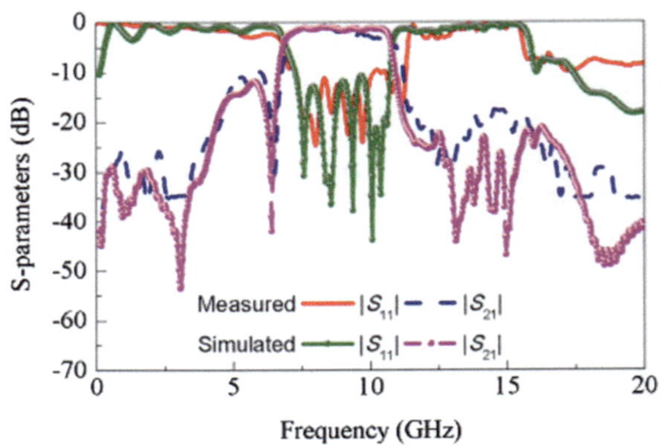

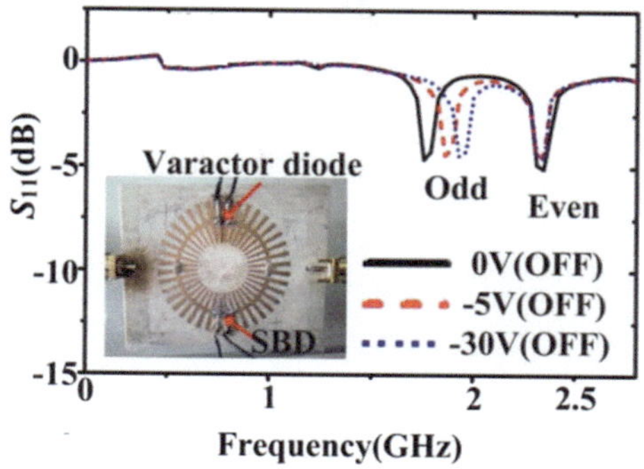

Fig. 4.3 S-parameter of spoof LSPs BPF with fabricated prototype in inset [4]

polariton T-shaped resonator is introduced in which the tuning can be achieved by using variable capacitor devices such as a varactor diode. The designed filter has 3-dB BW reconfigurability of 20 GHz (0.41–0.43 THz) and 300 MHz (100–400 MHz) with an insertion loss of ~3 dB and ~1.7–3.2 dB, respectively. A planar SSPP ring resonator is used to design a SSPP-based reconfigurable bandpass filter [2, 3]. These structures can be used to have multiple pass bands at different frequencies in a pre-defined frequency range. Further varactor diodes can be used with their resonators to achieve tunable passbands. A spoof localized surface plasmon (LSPs) can be used to design a resonating structure to realize a bandpass filter [4]. A double slit in a corrugated ring is used to control the spoof LSPs propagating in a resonator shown in Fig. 4.3. In this structure, a combination of Schottky barrier diode and varactor diode is used. The passband can be switched ON- and OFF by the Schottky diode whereas the pass band can be electronically tuned by the varactor diode. The resonant frequencies vary from 1.75 GHz to 1.85 GHz and 1.92 GHz corresponding to the bias voltages 0 V, −5 V, and −30 V, respectively.

Apart from ring resonators or spoof LSPs, graphene also constitutes an important place for tuning circuits. A tunable graphene filtering attenuator, namely the graphene-based effective SSPP (ESSPP) attenuator can be effectively used [5]. It consists of an ESSPP waveguide and a graphene sandwich structure (GSS). The effect of GSS on the attenuation and dimensions of the ESSPP waveguide on the passband shows that not only the attenuation of this attenuator can be tuned but also that the upper and lower cut-off frequencies can be adjusted independently (Fig. 4.4).

In this case, a combination of the p-i-n diode and varactor diode is used to achieve switching and tuning, respectively [6]. A continuously tunable cut-off frequency filter is composed of tunable spoof surface plasmon polaritons (SSPPs) loaded with p-i-n diodes and varactor diodes. The SSPP unit cell with the p-i-n diode can accomplish the reconfigurability at two different cut-off frequencies of 6.51 and 7.68 GHz. To realize more tunable cut-off frequencies, a new unit structure that can obtain a cascaded combination of varactor diodes and p-i-n diodes is presented. The filter can be tuned continuously from 5.34 to 7.23 GHz by controlling the magnitude of the biasing voltage, and the relative tunable bandwidth of the SSPP filter is up to 70.8%.

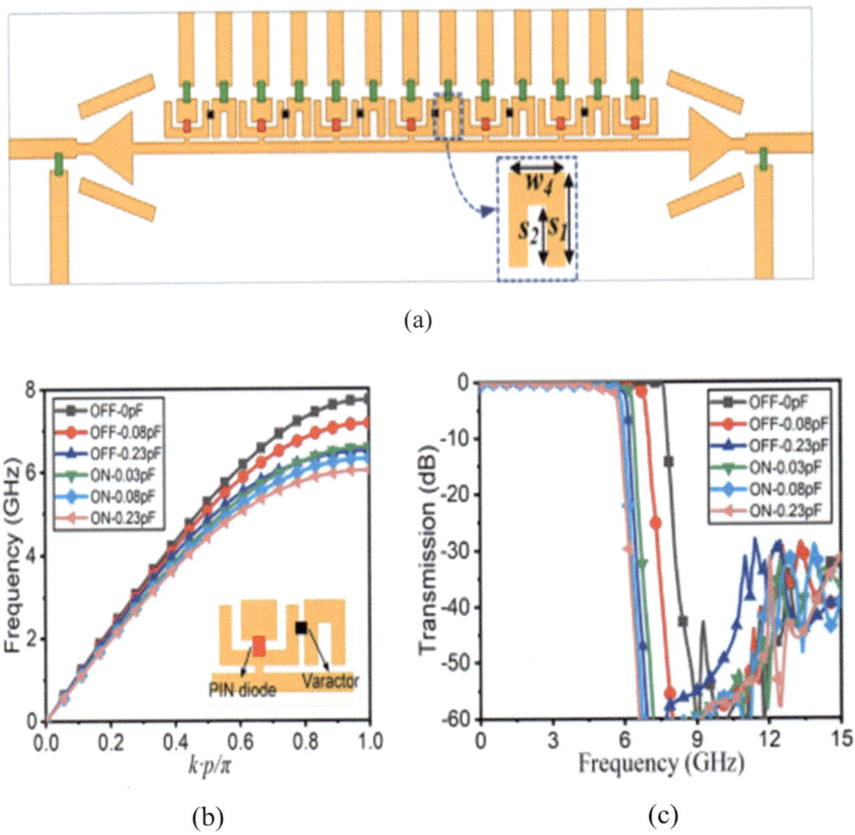

Fig. 4.4 **a** Schematic of tunable filter **b** dispersion curve of SSPP **c** simulated transmission coefficients [6]. Reproduced with permission from the IEEE

In the frequency range of 6–10 GHz, the transmission characteristic of a compact bandpass filter (BPF) with a reconfigurable rejection based on spoof surface plasmon polaritons (SSPPs) and an M-Cusp-H-Type (MCHT) cell shows a harmonics suppression level above 30 dB [7]. Moreover, by using varactor diodes, lumped elements, and direct current (DC) bias circuits, the designed BPF has a low cut-off frequency and reconfigurable rejection. The designed BPF allows the stopband's rejection level, frequency, and bandwidth to be tuned by changes in the external bias voltage. When combined with varactor diodes, the designed BPF displays a stable passband between 3.45 and 4.91 GHz in frequency. With this configuration, wideband transmission suppression is impressive, exceeding 30 dB starting at 6–10 GHz.

4.3 Tunable Band-Notched Filter

Band-notched filter with fixed notched frequency is already discussed in the previous chapter. In addition to these fixed notch band filters, reconfigurable notch filters form a very important concept for adaptive applications such as IoT where our circuit works for multiple applications simultaneously. Figure 4.5 shows a high permittivity BST dielectric block placed in the gap of an SSPP transmission line [8]. The notch frequency is sensitive to the variation of the dielectric constant of the BST ceramic, a transmission dip up to −11.5 dB is observed at the notch frequency of 8.12 GHz when the dielectric constant of BST is 425; the notch frequency shifts from 8.7 to 7.6 GHz when the dielectric constant of BST increases from 375 to 475. In [9], a notch band has been created after getting to couple between the split ring resonator and the transmission line, again it can be tuned in the range of 3.3–4.5 GHz. A similar idea is reported in [10] by loading half-wavelength SPP resonators along the weaker field confinement side of the asymmetrical SPP waveguide; dual notches can be generated in the SPP transmission.

Varactor diodes, with voltage-controllable capacitance, are incorporated into the SPP resonators to obtain dual and tunable notches in the frequency range of 3.765–4.29 GHz and 4.85–5.69 GHz, respectively. Rejection bands are achieved in [11] after mounting the Schottky diode or varactor diode across the slit in a C-shaped ring loaded on the back of the transmission line, the obtained notch band was tuned from 5.8 to 6.18 GHz. When the SSPP waveguide is loaded with a half-wavelength slot resonator on both sides, a notch is introduced in the passband [12]. Meanwhile, the SSPP waveguide with two ring slot resonators is evaluated, whose results are consistent with the SSPP low-pass filter. Finally, two PIN diodes are bridged across the slits of the two resonators to realize the dynamic switch ability of the notch. The waveguide functions as a notch filter when the diodes are in ON-state and a low-pass filter when the diodes are in OFF-state.

A PIN diode is integrated into the SRR of the notch filter to make the notch frequency switchable [13]. With periodic arrays of ultra-thin metallic meander-strip units, a novel range of spoof surface plasmon polariton (SSPP) waveguides and notch filters is presented for strong confinement of microwave SSPP transmission and notch rejection with high efficiency. The asymptotic frequency of the proposed waveguide reduces to nearly half as that of the SSPP waveguide based on traditional rectangular-grooved units with the same transverse size. A notch filter is designed by loading a split ring resonator (SRR). The transmission coefficient (S_{21}) of this filter is −18 dB at the notch frequency of 2.52 GHz. As the PIN diode turns from the ON-state to the OFF-state, the notch frequency switches from 2.5 to 3.5 GHz.

Spoof surface plasmon polaritons (SPPs) of hollowed-bowtie (HBT) cells are used to design a tunable band rejection filter [14]. Direct current (DC) bias circuits, varactor diodes, and lumped elements work together to enable us to design a bandpass filter with a low cut-off frequency and reconfigurable rejection. The rejection level, frequency, and bandwidth of the stopband are adjusted by the bias voltage regulation. A ground plane is very important for reconfigurable circuits as reconfigurability is

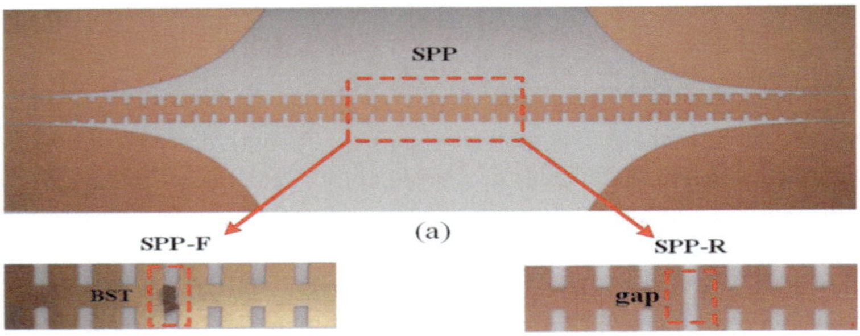

(a)

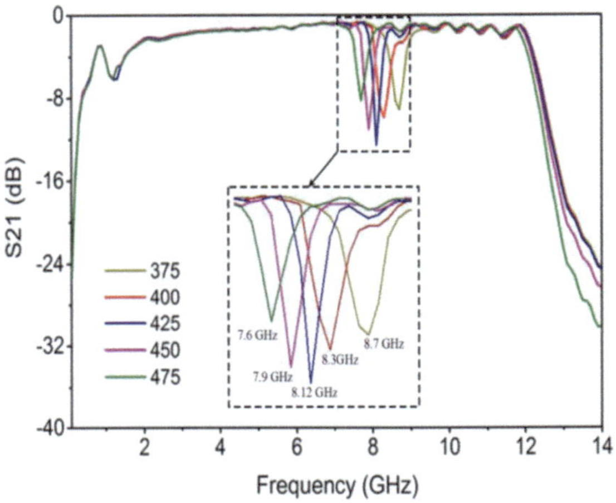

(b)

Fig. 4.5 **a** Photograph **b** results of tunable band-notched SSPP transmission line [8]

achieved by integrating diodes and they require a bias circuit so grounding plays an important role. Spoof surface plasmon polariton (SSPP) waveguides based on double-sided parallel-strip lines (DSPSL) are formed by etching opposite spiral grooves on the upper and lower metal layers of the DSPSL as shown in [15]. Compared with the conventional comb-shaped SSPP structures at the same asymptotic frequency, the size of the structure is reduced to 19.85%, which facilitates miniaturization and integration of microwave devices. Two-notch filter based on the proposed SSPP waveguide can achieve adjustable narrowband or broadband filtering by loading different capacitors.

A band rejection filter with dynamic modulation based on the spoof surface plasmon polaritons (SSPPs) waveguide is reported in [16]. The primary source of dynamic adjustment is the alteration of the capacitance between the waveguide configuration and the U-shape as shown in Fig. 4.6. The capacitance can alter the fundamental mode cut-off frequency. The high-order propagation mode and the

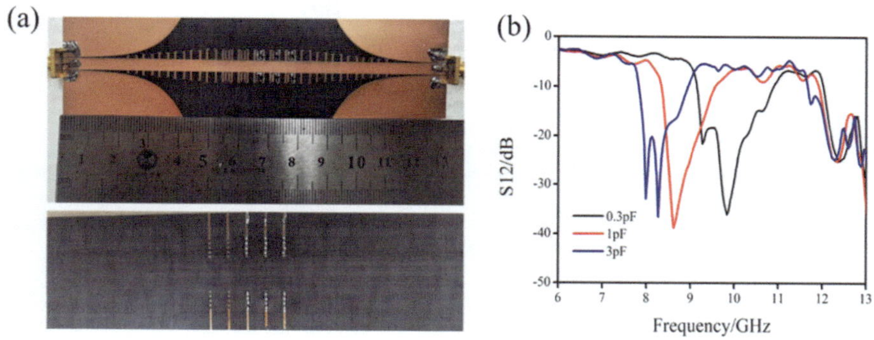

Fig. 4.6 **a** Photograph of fabricated structure **b** measured S-parameter under different capacitances [16]

fundamental mode's cut-off frequency combine to form the rejection band. Around 2 GHz, real-time modulation was accomplished by varying the direct voltage applied to the varactor diode.

4.4 Design Flow of Tunable Band-Notched Filter from SSPP Transmission Line

In this section, we will discuss the design flow of a tunable band-notched filter using an SSPP transmission line. We will discuss in detail the effect of incorporating a slot and then how tuning can be achieved by varying the capacitance as well as slot width. Next, the circuit analysis of the structure will be discussed and finally, the results will be concluded with the experimental validation [17].

4.4.1 Incorporation of Slot Resonator into SSPP Transmission Line to Obtain Band-Notched Characteristics

To obtain a band-notched characteristic, a slot resonator is incorporated into the SSPP transmission line as shown in Fig. 4.7. Here, two slots of different sizes are created to get two-notch bands at different frequencies. The width and length of Slot 1 are w_1 and l_4, respectively whereas the width and length of Slot 2 are w_3 and l_4, respectively. After inserting a slot resonator loaded with a capacitor, a notch is created in the low-pass characteristic of the SSPP-TL. The center frequency of the notch band can be tuned from 3.5 to 10.5 GHz by changing the loading effect (capacitor). Similarly, another slot resonator of a different size loaded with a capacitor is incorporated, which creates another notch band at a different frequency. By controlling the slot

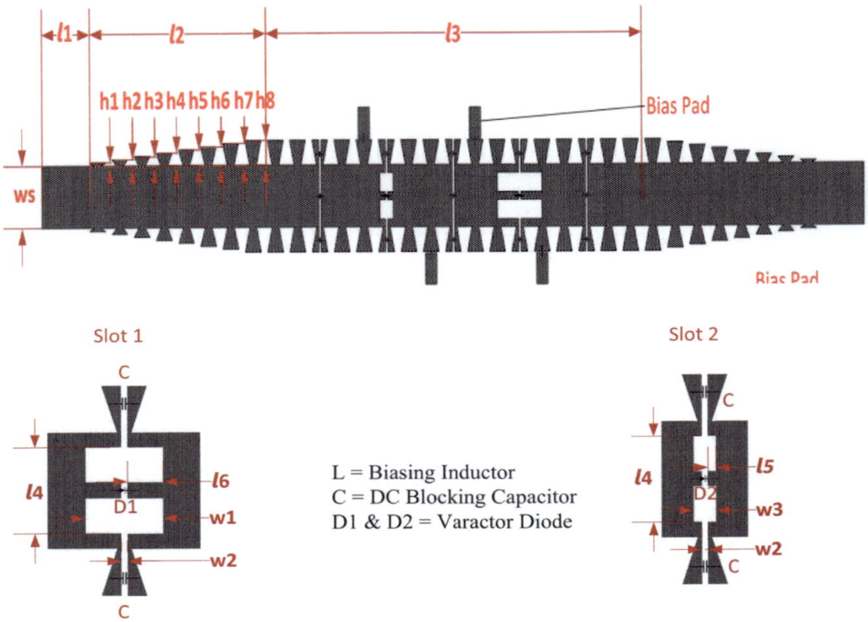

Fig. 4.7 Notch filter incorporated into the SSPP transmission line; slot 1 and slot 2 are loaded to the line to get single or/and dual notches from the structures

resonator's width or the capacitor value, single or/and dual-notch band characteristics can be achieved.

For demonstration purposes, to get tunable notches, capacitors are replaced with varactor diodes. D_1 and D_2 are the varactor diodes as shown in Fig. 4.7. The small slits of width ($w_2 = 0.2$ mm) are cut on the line for separating the bias and additional capacitors (C) of 22 pF are placed across the slits to block direct current (DC) while inductors (L) of 100 nH are incorporated to properly bias the varactor diodes. When the width of both the slots is the same, i.e., $w_1 = w_3$, a single notch is obtained but it can be tuned by varying the capacitance (through varactor) across the slot. Various combinations of slots 1 and 2 have been analyzed. Initially, the widths of the slots were kept the same and constant, but the capacitance across the slots varied by the same amount. When the widths (w_1, w_3) of the slots were kept 1 mm and varied the capacitances across the slots from 0.2 to 1 pF, the center frequency of the notch varied from 10.5 to 5 GHz. Similarly, at $w_1 = w_3 = 2$ mm, when capacitance was varied from 0.2 to 1 pF, the notch was varied from 8 to 3.9 GHz. The tuning of the notch is analyzed for different widths of slots, but here we are showing only three values ($w_1 = w_3 = 1$, 2, and 3) as shown in Fig. 4.8a–c. It is observed from the above analysis that the center frequency of the notch reduces with an increase in the slot width. It is further verified with the help of circuit model analysis discussed in Sect. 4.4.2.

Fig. 4.8 Simulated transmission parameters of the slot-loaded SSPP transmission line for **a** slot width, $w_1 = w_3 = 1$ mm **b** slot width, $w_1 = w_3 = 2$ mm **c** slot width, $w_1 = w_3 = 3$ mm and **d–f** simulated transmission parameters for a different combination of capacitance values for slot 1 and slot 2 with $w_1 = 1.2$ mm, and $w_3 = 3.6$ mm

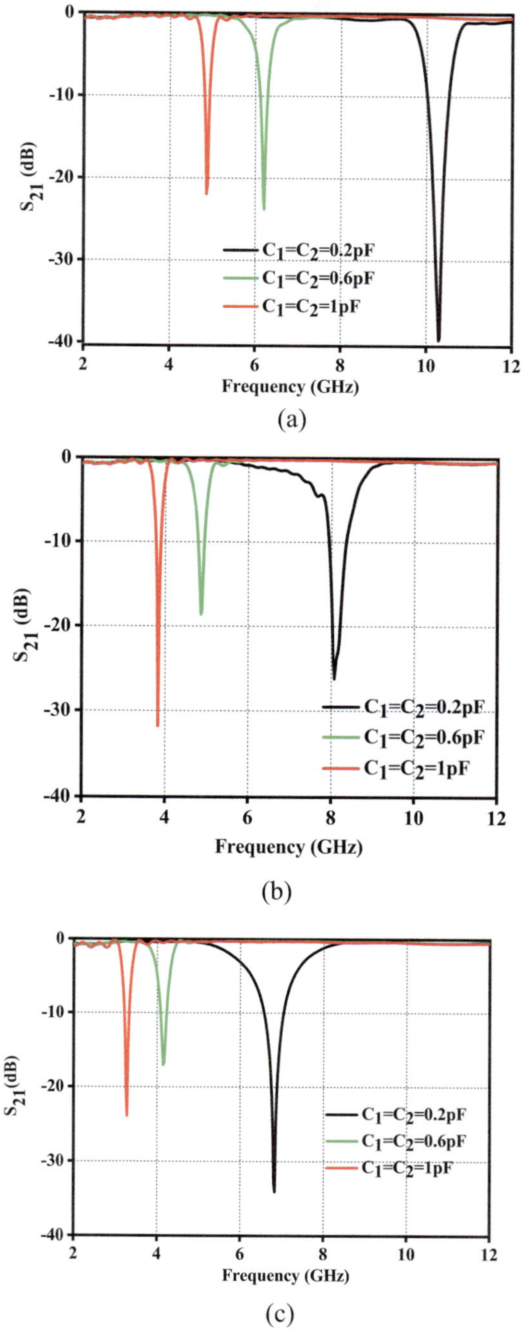

Fig. 4.8 (continued)

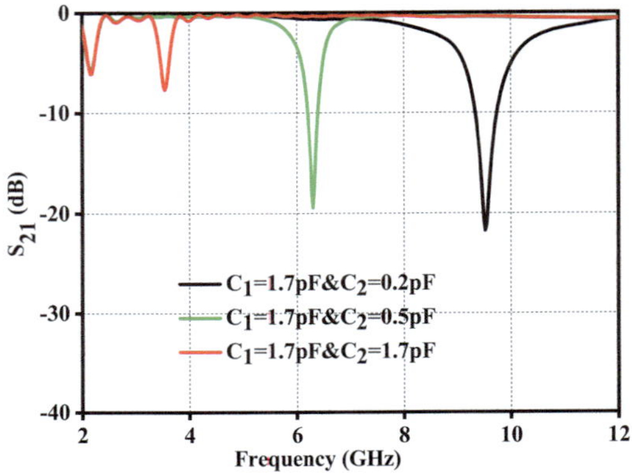

(d)

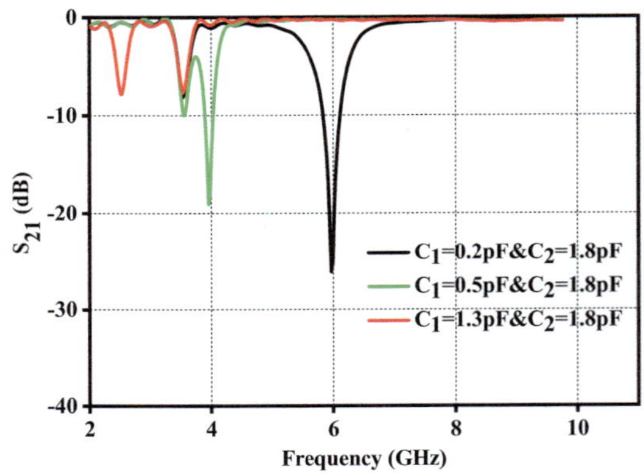

(e)

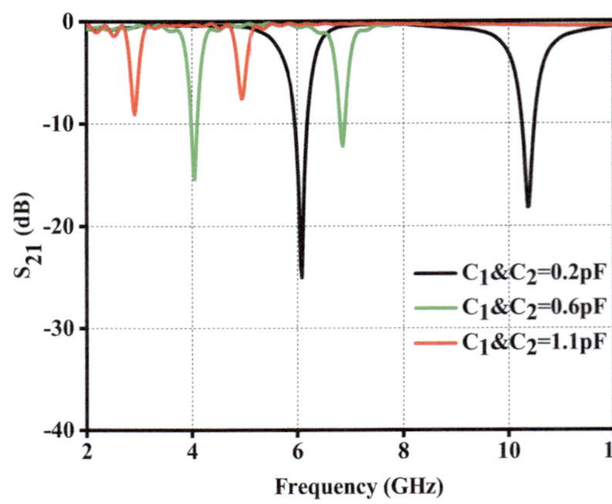

(f)

Next, the slot widths were varied. Again, various combinations were considered for getting appropriate behavior of the line and analyzing the S-parameters. With optimized values of both slots ($w_1 = 1.2$ mm, and $w_3 = 3.6$ mm), a single notch having a tuning band from 3.5 to 11 GHz is obtained. Additionally, the dual-notch bands are obtained by changing the values of capacitances across the slots. Figure 4.8d–f shows the simulated transmission parameter by considering a different value of capacitances across both slots. Figure 4.8d plots the transmission coefficients when the capacitance C_1 is kept constant at 1.7 pF while C_2 is varied from 0.2 to 1.7 pF, the notch varies from 9.5 to 3.5 GHz. Here, C_1 and C_2 are the capacitance for slot 1 and slot 2, respectively, that are used to carry out the simulations. During measurements, C_1 corresponds to varactor D_1 while C_2 corresponds to varactor D_2.

The transmission coefficient is below -10 dB when C_2 is in the range of 0.2–0.7 pF. Similarly, Fig. 4.8e plots the transmission coefficients when the capacitance C_2 is kept constant at 1.8 pF while C_1 is varied from 0.2 to 1.3 pF; in this case, the notch is varied from 6 to 2.5 GHz. Figure 4.8f shows plots when C_1 and C_2 are varied by the same amount from 0.2 to 0.8 pF, dual notches are obtained, and the first notch is varied in the range of 6–10.5 GHz while the other one is varied in the range of 3.8–6.2 GHz. In this way, single or/and dual notches are obtained from the transmission line with a wide tuning range. In practice, two varactor diodes (MA46H120) namely D_1 and D_2 are employed across the slots to achieve an electronically tunable notch. It can be concluded from the above study that the notch can be tuned from 3.5 to 9.5 GHz by varying the capacitances across the slots.

4.4.2 Circuit Model Analysis of Slot-Loaded SSPP Transmission Line

Figure 4.9a shows the equivalent circuit of the trapezoidal unit cell. L_{eq1}, L_{eq2}, and L_{eq3} are the inductances caused by the current flowing on the metal surface and it can be calculated from (4.1). The circuit inductance can be calculated using the formula

$$L_i = \frac{\mu_o l_i}{2\pi}\left(\ln\frac{2l_i}{w_i + t} + \frac{1}{2}\right), \tag{4.1}$$

where 'l_i' the length, 'w_i' is the width, and 't' is the thickness of the unit cell. The equivalent inductance is affected by varying the width and length of the unit cell, but the thickness of the unit cell does not have much impact on the equivalent inductance as it is very small compared to the width or length of the unit cell. The equivalent inductance increases with an increase in the length or a decrease in the width of the unit cell. C_{eq1} is the capacitance between the pair of metallic tracks.

Its value increases with an increase in the tracks (metal part) area or a decrease in the gap between two tracks. This capacitance is inversely proportional to the asymptotic cut-off frequency. C_{eq2} models the leaky currents from the transmission

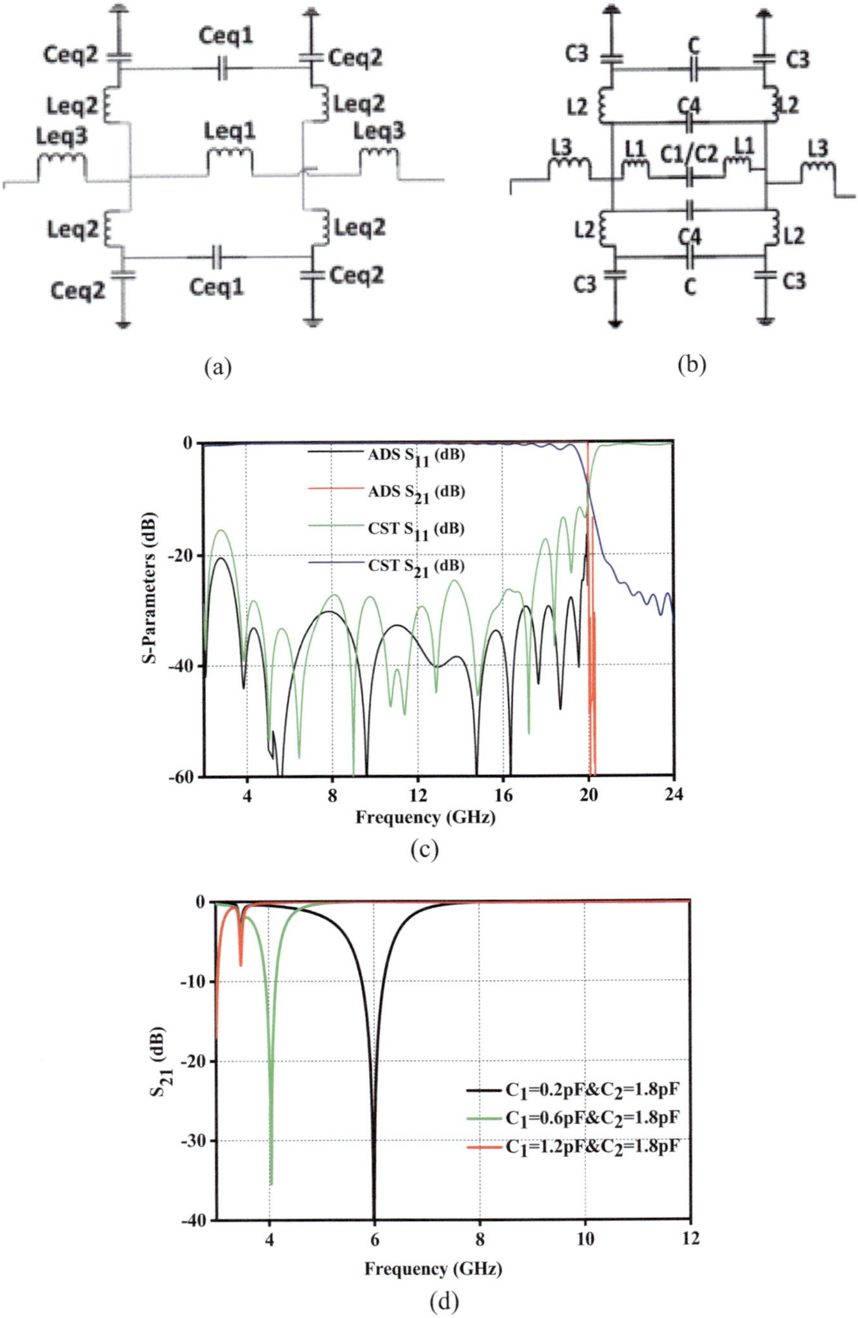

Fig. 4.9 **a** Equivalent circuit model of the unit cell **b** equivalent circuit model of the slot-loaded unit cell **c** S-parameters of the SSPP transmission line **d–f** transmission coefficients of the slot-loaded SSPP transmission line for a different combination of capacitances across slot 1 and slot 2

Fig. 4.9 (continued)

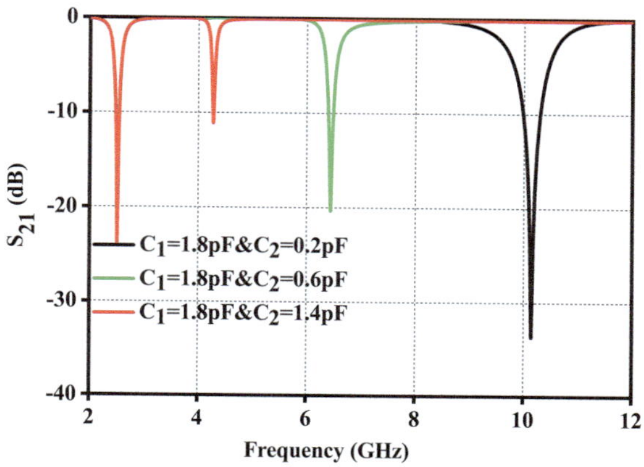

(e)

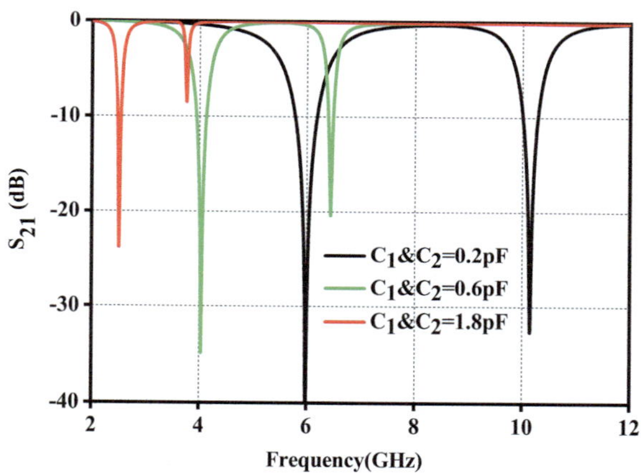

(f)

line to the ground plane like the leaky currents in the microstrip line. C_{eq2} increases with an increase in the height of the corrugation or a decrease in the width of the corrugation. The calculated values of inductances and capacitances are listed in Table 4.1 [18]. Figure 4.9b shows the equivalent circuit of the slot-loaded unit cell. The capacitance C_4 arises after the insertion of a slot in the unit cell, C_1 (or C_2) is the capacitance of the varactor diode D_1 (or D_2), L_1 is the inductance that arises due to the metallic lines present in the center of the slot, and C is the dc blocking capacitor.

The equivalent circuits of the unit cell and slot-loaded unit cell are simulated with the help of Agilent Advanced Design System (ADS). S-parameters of the equivalent circuit are plotted and compared with those obtained from CST Microwave Studio as shown in Fig. 4.9c. Figure 4.9d–f shows the transmission parameter of the line after varying the capacitances C_1 and C_2. Single notch or/and dual notches are obtained from the circuit model analysis of the slot-loaded unit cell.

Table 4.1 Calculated values of the components for the equivalent circuit of unit cell

	$h = 0.2$ mm	$h = 0.6$ mm	$h = 1.2$ mm	$h = 1.8$ mm
L_{eq1} (nH)	0.2603	0.2603	0.2603	0.2603
L_{eq2} (nH)	0.080	0.100	0.170	0.294
L_{eq3} (nH)	0.360	0.360	0.360	0.360
C_{eq1} (pF)	0.008	0.018	0.041	0.060
C_{eq2} (pF)	0.0012	0.019	0.0025	0.0098

4.4.3 Experimental Validation of the Concept

The slot-loaded SSPP transmission line is fabricated to validate the proposed concept. Figure 4.10a shows the photograph of the fabricated line and Fig. 4.10b shows the measurement setup. The transmission coefficients of the transmission line are measured using a vector network analyzer (Model: PNA-E8364C). Figure 4.10c–e shows the measured transmission coefficients of the line considering different bias conditions across the varactor diodes.

When the bias voltage applied across the varactor diode (D_1) is 0 V (fixed) while the bias voltage across D_2 is varied from 0 to 19 V, and the notch varies from 6.5 to 10.5 GHz. At the notch frequencies, S_{21} (transmission coefficient) is below -10 dB as shown in Fig. 4.10c. Similarly, when the bias voltage across D_2 is kept fixed at 0 V while the bias voltage across D_1 is varied from 0 to 18 V, the center frequency of the notch is varied from 3.5 to 6.5 GHz. Again, the value of S_{21} is below -10 dB over the notch band. It can be concluded from the above observation that the notch band can be tuned from 3.5 to 10.5 GHz by varying the bias voltage across D_1 and D_2. When the bias voltage across D_1 and D_2 are varied by the same amount, two notches are obtained, the first notch varies from 3.9 to 6.5 GHz while the second notch varies from 6 to 10.5 GHz.

By considering a different combination of bias voltages applied across varactors, single notch or/and dual notches are obtained from the proposed slot-loaded line. Compared to the previously reported band-notched transmission lines, the proposed structure provides a wide tuning range of a single notch or/and dual notches with smaller number of diodes. The specialty of the proposed work is that it covers the full operating band of UWB systems. There is a minute difference in the simulated and measured results. Firstly, the variation in the frequency is due to the deviation in the actual value of the capacitor given in the datasheet and used in the simulations. Secondly, the broadening of the notch band is due to the insertion loss in the varactor diode and the use of not very high-quality capacitors.

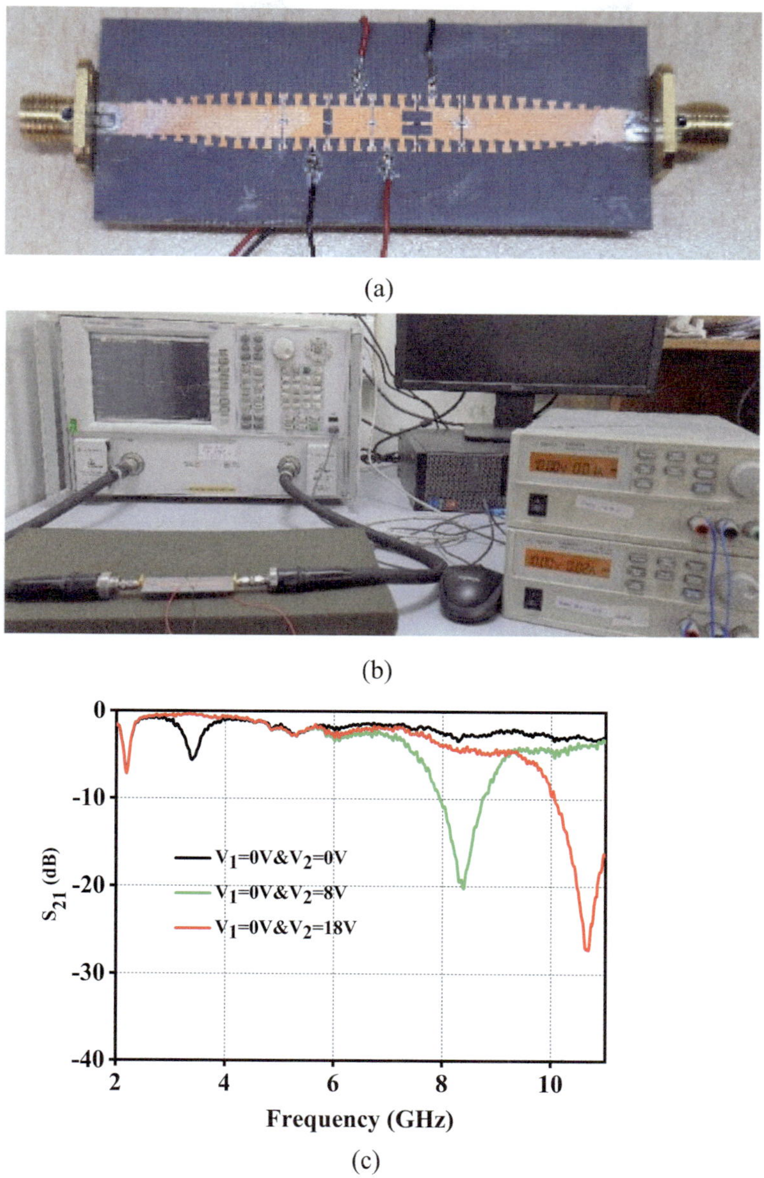

Fig. 4.10 **a** Fabricated prototype of the slot-loaded SSPP transmission line **b** measurement set up **c–e** measured transmission coefficients of the slot-loaded transmission line considering different bias voltages across varactor diodes

Fig. 4.10 (continued)

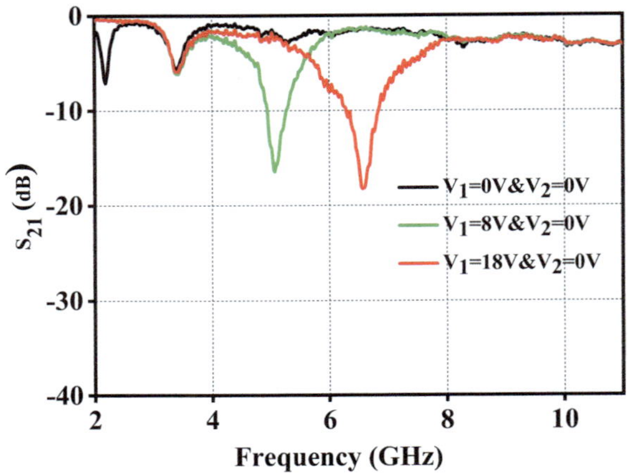

(d)

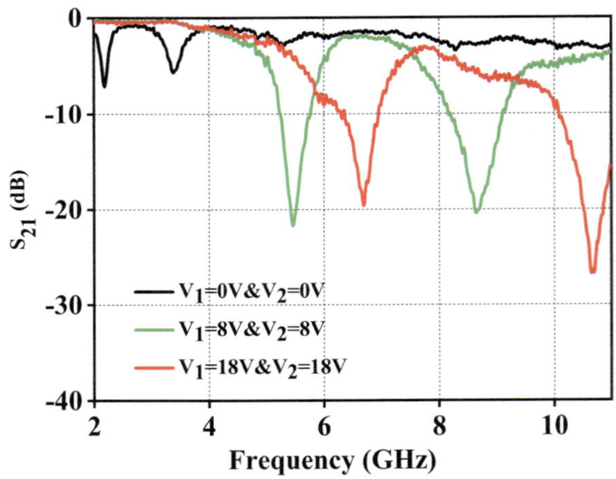

(e)

4.5 Conclusion

In this chapter, different techniques to design a tunable filter are discussed. We conclude that diodes such as the varactor diode, Schottky diode, and p-i-n diodes can be used to achieve tunability. Apart from diode, graphene and high permittivity substrates can also be used to achieve tuning. Different techniques to design bandpass and band-notched structures using SSPP are discussed in detail. It is observed with the use of SSPP, wide tuning can be achieved. Further detailed analysis of band-notched structures is discussed in this chapter. It discusses two aspects one is to reduce the impact of interfering signals existing in the ultra-wideband spectrum and the second is to minimize emission in out-of-band frequencies to protect other circuits or components against interference. A single tunable notch or two independently tunable notches have been achieved to block interfering signals in the ultra-wideband frequency range. A single tunable notch was obtained in the frequency range of 3.5–10.5 GHz by employing varactor diodes. Two independent tunable notches have also been achieved by applying a different combination of bias voltages across varactor diodes. The proposed idea of creating a single notch and dual notches in the pass band behavior of the SSPP line covers the full operating range of the UWB system. The performance of the slot-loaded SSPP line with reconfigurable notches has been analyzed and verified after comparing the results obtained through equivalent circuit models and measurements.

References

1. Wang, J., Zhao, L., Hao, Z.-C.: A band-pass filter based on the spoof surface plasmon polaritons and CPW-based coupling structure. IEEE Access **7**, 35089–35096 (2019). https://doi.org/10.1109/ACCESS.2019.2903147
2. Jaiswal, R.K., Pandit, N., Pathak, N.P.: Spoof surface plasmon polaritons based reconfigurable band-pass filter. IEEE Photonics Technol. Lett. **31**(3), 218–221 (2019)
3. Jaiswal, R.K., Pandit, N., Pathak, N.P.: Spoof surface plasmon polariton-based reconfigurable band-pass filter using planar ring resonator. Plasmonics **14**, 631–646 (2019). https://doi.org/10.1007/s11468-018-0841-0
4. Zhou, Y.J., Zhang, C., Yang, L., Xiao, Q.X.: Electronically switchable and tunable bandpass filters based on spoof localized surface plasmons. J. Opt. Soc. Am. B **34**, D9–D12 (2017)
5. Yi, Y., Zhang, A.-Q.: A tunable graphene filtering attenuator based on effective spoof surface plasmon polariton waveguide. IEEE Trans. Microw. Theory Tech. **68**(12), 5169–5177 (2020)
6. Yang, H., et al.: Spoof surface plasmon polariton-based continuously tunable cutoff frequency filter. IEEE Microw. Wirel. Technol. Lett. **33**(10), 1430–1433 (2023)
7. Liu, H., Cheng, Y., Luo, H., Chen, F., Yang, L., Li, X.: Reconfigurable microstrip bandpass filter based on spoof surface plasmon polaritons of M-Cusp-H-Type cells. IEEE Trans. Circuits Syst. II Express Briefs **71**(6), 3031–3035 (2024)
8. Luo, J., He, J., Apriyana, A., Feng, G., Huang, Q., Zhang, Y.P.: Tunable surface-plasmon-polariton filter constructed by corrugated metallic line and high permittivity material. IEEE Access **6**, 10358–10364 (2018)
9. Xu, J., Zhang, H.C., Tang, W.X., Guo, J., Qian, C., Li, W.Y.: Transmission-spectrum-controllable spoof surface plasmon polaritons using tunable metamaterial particles. Appl. Phys. Lett. **108**, Art. no. 191906 (2016)
10. Le Zhang, Q., Chan, C.H.: Spoof surface plasmon polariton filter with reconfigurable dual and non-linear notched characteristics. IEEE Trans. Circuits Syst. II: Express Briefs **68**(8), 2815–2819 (2021)
11. Zhou, Y.J., Xiao, Q.X.: Electronically controlled rejections of spoof surface plasmons polaritons. J. Appl. Phys. **121**, Art. no. 123109 (2017)
12. Li, J., Xu, K.-D., Shi, J., Guo, Y.-J., Zhang, A.: Spoof surface plasmon polariton waveguide with switchable notched band. IEEE Photonics Technol. Lett. **33**(20), 1147–1150 (2021)
13. Ye, L., Chen, Y., Wang, Z., Zhu, C., Zhuo, J., Liu, Q.H.: Compact spoof surface plasmon polariton waveguides and notch filters based on meander-strip units. IEEE Photonics Technol. Lett. **33**(3), 135–138 (2021)
14. Pan, L., Wu, Y., Wang, W.: Bandpass filter with reconfigurable rejection and deep-upper-wideband harmonics suppression using spoof surface plasmon polaritons of hollowed-bow-tie cells. IEEE Trans. Microw. Theory Tech. **71**(4), 1826–1834 (2023)

15. Ye, L., et al.: J. Phys. D Appl. Phys. **53**, 265502 (2020)
16. Yang, W., Liu, L., Su, X., Dong, L., Liu, Y., Li, Z.: Dynamic modulation band rejection filter based on spoof surface plasmon polaritons. Electronics **9**, 993 (2020)
17. Sharma, S., Singh, R.K., Basu, A., Koul, S.K.: Reduction of in-band interfering signals and minimize emission in out-of-band frequencies for EMI/EMC applications. IEEE Trans. Electromagn. Compat. https://doi.org/10.1109/TEMC.2024.3472795
18. Paul, C.R.: Analysis of Multiconductor Transmission Lines, 2nd ed. (2008)

Chapter 5
Frequency Scanning Antennas Using Spoof Surface Plasmon Polariton

5.1 Introduction

Frequency scanning antenna also known as leaky-wave antenna is a type of traveling wave antenna having outstanding properties such as wide frequency scanning capability, high directivity, and simple feed network [1]. Conventional guiding structures such as microstrip lines and waveguides have low radiation in the broadside direction and create a problem in scanning a beam through the broadside due to the limitation of narrow beam scanning and open stop band (OSB) region. An open stop band region is created due to the coupling of two oppositely directed space harmonics with equal amplitudes, causing reflection that limits the radiation in the broadside direction. OSB can be eliminated by using different techniques such as asymmetrical loading of radiating elements, reflection cancelation, CRLH, and impedance matching structures. The electrical length between the adjacent radiating element is adjusted to get a proper phase delay to minimize the OSB has been discussed in [2]. Transmission lines loaded with capacitive gaps and stubs are also used to eliminate OSB. Asymmetrical structures are realized to eliminate the open stop band region [3–7]. LWA was designed using microstrip step discontinuities, insertion of vias helped in eliminating the OSB region, and forward-to-backward scanning was achieved from the structure [8]. Substrate integrated waveguide-based LWAs have also been reported to achieve a wide forward-to-backward scanning by eliminating the OSB but these LWAs have low radiation efficiency and narrow bandwidth [9–13]. The SSPP-TLs used in the realization of LWAs have gained attention due to their high field confinement, low loss, and low profile [14, 15]. In SSPP-TL-based LWA, the technique used to eliminate the OSB is asymmetrical placement of radiating element which use the basic concept of phase delay. To suppress the open stopband, circular patches are placed in decreasing size with a multiplication factor of 0.7 and with good impedance matching reflects the suppression of the open stopband [16]. To mitigate the open stopband problem, the width of the metal groove on the top layer and the bottom layer are made asymmetrical [17]. The center of the open-loop element is misaligned

S. K. Koul and S. Sharma, *Microwave Integrated Circuits Based on Spoof Surface Plasmon Polariton*, Signals and Communication Technology,
https://doi.org/10.1007/978-981-96-9041-1_5

with the center of the SSPP unit cell, which can increase the asymmetry of the leaky-wave antenna to suppress the open stopband (OSB) effect, which is characteristic of periodic leaky-wave antennas [18]. By introducing the unique dual-half-elliptical radiation patches in the LWA design, the radiation efficiency is increased and the reflection is reduced simultaneously in the operating band to mitigate the open stop-band (OSB) effect [19]. Spoof surface plasmon polariton-based leaky-wave antenna plays an important role in the field of research and development and shows promising results compared to the conventional transmission line.

5.2 Frequency Scanning Antenna

The leaky-wave antennas can be classified into two types according to the feeding mechanism, i.e., traveling wave antenna and standing wave antenna.

5.2.1 Traveling Wave Antenna

A class of antennas known as traveling wave antennas uses a traveling wave on a guiding structure as its primary radiating mechanism. Traveling wave antennas can be broadly classified as either fast wave or slow wave, also known as leaky-wave antennas. The guided wave in a slow wave antenna propagates at a phase velocity lower than the speed of light in free space, which is referred to as a slow wave. Radiation only happens at discontinuities, and such a wave does not intrinsically radiate. On the other hand, a wave on a leaky-wave antenna (LWA) might be fast, moving at a phase velocity faster than light. Since the propagation wavenumber of this kind of wave is complex and consists of both a phase and an attenuation constant, it radiates continuously along its length. With this kind of antenna, low sidelobe levels and highly-directive beams can be produced at any given angle. Depending on the kind of guiding structure, leaky-wave antennas fall into two major categories: periodic and uniform. A waveguide that has been partially opened to allow radiation to occur is typically the form of a uniform structure, with a cross-section that is uniform (constant) along the length of the structure. Since the guided wave on the uniform structure moves quickly, it radiates as it does so. An antenna structure with periodic leaky-wave characteristics is composed of a uniform structure supporting a slow, non-radiating wave that has undergone periodic modulation. Since slow waves radiate at discontinuities, the wave continuously radiates along the length of the structure due to periodic modulations (discontinuities). The SSPP-based LWA are usually periodic structures as they work on the concept of the slow wave.

5.2.1.1 Standing Wave Antenna

Standing wave antenna as the name suggests is working on the principle of standing waves. Standing waves can be generated by the superposition of two opposing traveling waves. When the traveling wave is repeatedly bounced back in a guided wave structures to exactly overlap each other, standing wave is generated. These can be generated by using open ends, short ends, and impedance mismatch nodes at the end opposite to the feeding point. The degree of this exactness determines the guide standing wave ratio. This value ranges between 1 and infinity, with 1 representing a pure traveling wave and infinity representing a pure standing wave.

5.3 Leaky-Wave Antenna Using Spoof Surface Plasmon Polariton

Groundless SSPP-TLs are used in developing LWAs. In [20–23], the SSPP LWAs are realized using an array of circular patches as radiators that convert a slow wave mode (SSPP mode) into a fast wave mode (radiating mode) to achieve good radiation efficiency. Different types of SSPP-TLs are used to couple RF power to the radiating element. The geometry of the leaky-wave open-loop antenna array fed by SSPPs is shown in Fig. 5.1 [23]. It features radiation efficiency exceeding 70%, stable gain of approximately 13 dBi, and wide impedance bandwidths. Figure 5.2 displays the constructed prototypes of the suggested leaky-wave open-loop arrays fed by SSPPs, complete with connectors. Effective excitation of the one-wavelength and one and a half wavelength modes of the open-loop can produce radiation fields with orthogonal polarizations when a straight SSPPs section near the open end of the loop is used as the feed. As shown in Fig. 5.3, the measured and simulated $|S_{11}|$ and $|S_{21}|$ of the suggested leaky-wave open-loop antenna arrays fed by the SSPPs are in good agreement. For the two open-loop arrays, the measured $|S_{11}|$ and $|S_{21}|$ are both below -10 dB over the operating bands, confirming promising impedance matching characteristics. The suggested antenna arrays' frequency scanning performance in yoz-plane is shown in Fig. 5.4 along with simulation and measurement results.

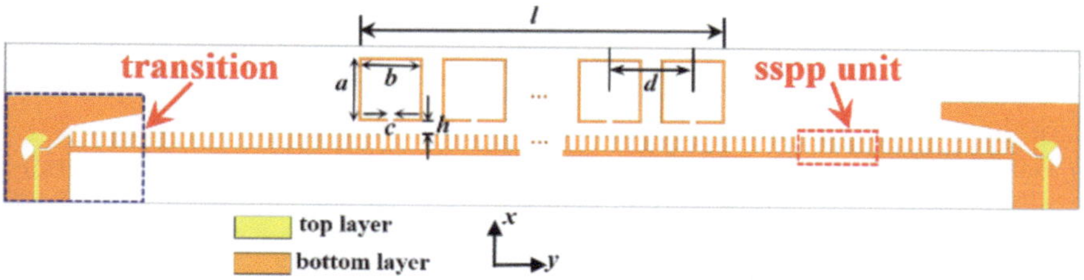

Fig. 5.1 Schematic of the proposed SSPP fed leaky-wave open-loop antenna array [23]

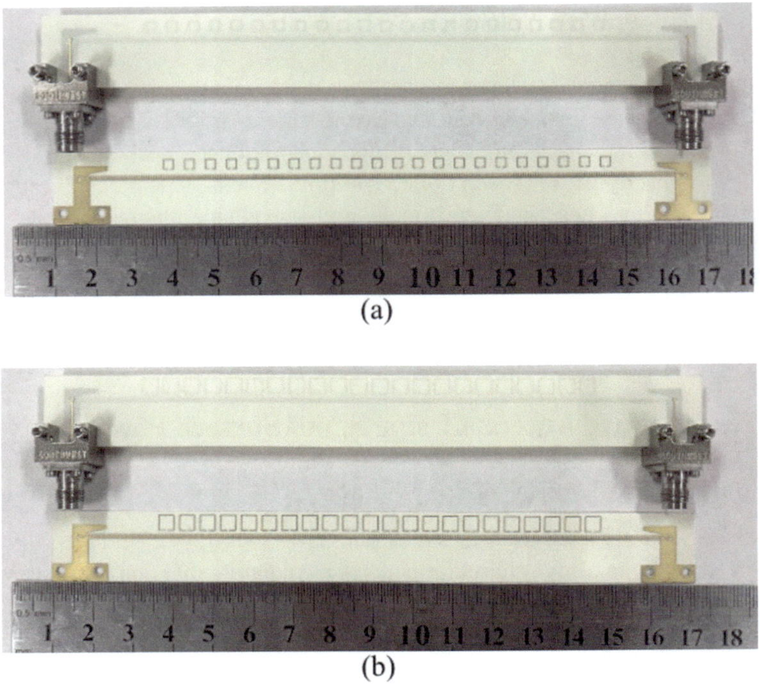

Fig. 5.2 Fabricated prototype of SSPPs-fed open-loop antenna **a** one-wavelength design, **b** one and a half wavelength design [23]

In [24], orthogonal rectangular patches periodically loaded on both sides of SSPP-TL are used to realize a circularly polarized LWA with a high radiation efficiency of 87% and a peak gain of 10.5 dBi but with a low scanning range of 15° (from −7° to +8°) across the operating band. Normally, groundless LWAs radiate dual beams due to the absence of a ground plane. Groundless LWA cannot be an optimum solution for applications such as sensing and imaging which require a directional beam. Grounded SSPP LWAs find application in long-range wireless networks, high-resolution sensing systems, and high-speed data communication devices. For long-range wireless networks, high-gain antennas are required. Additionally, there is a requirement for wide-angle scanning for various radar applications and radio astronomy. Although a groundless SSPP LWA was designed to radiate a single beam with a scanning range of 65°, it has significant radiation in the undesired directions [25].

In [26–30], grounded SSPP-TLs are used to realize LWAs. LWA with ground plane on the backside radiates a directional beam. In [26], PEC and artificial magnetic conductor (AMC) ground planes were used to achieve a directional beam and high gain with a scanning range of 43° but an air gap between the ground plane and antenna increases the complexity of the structure. In [27], the microstrip SSPP-TL along with a set of metallic posts arranged alternately along two sides of SSPPs-TL has achieved

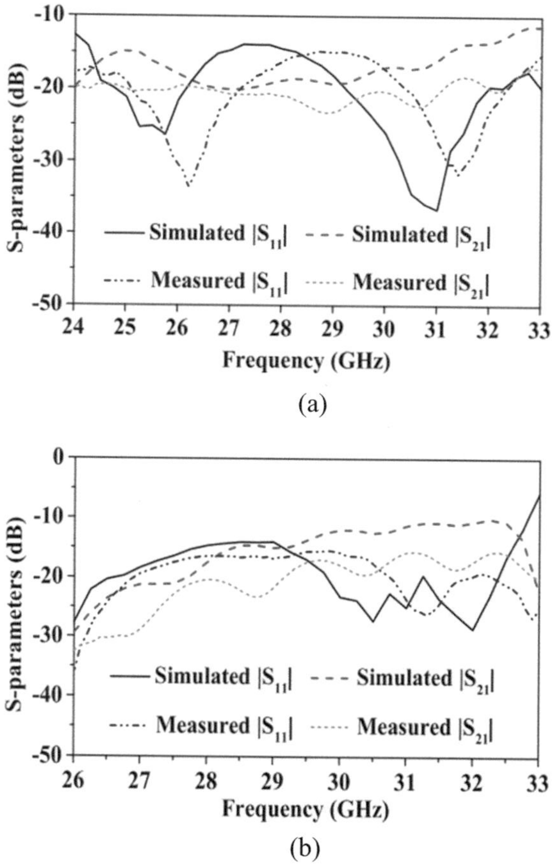

Fig. 5.3 Simulated and measured results of SSPPs-fed open-loop antenna **a** one—wavelength design, **b** one and a half wavelength design [23]

a wide scanning range but again the use of metallic posts has increased the fabrication complexity. In [28], the microstrip SSPP-TL was used to feed an array of patches to achieve circular polarization and wide-angle beam scanning of 66°. The antenna consists of two substrate layers; there may be a chance of misalignment between both layers, an air gap at the connector transition and screws near the feed may affect the performance of the antenna.

The plasmonic waveguide model with sinusoidal modulation is illustrated in Fig. 5.5 [31]. Figure 5.6a shows the artificially created sinusoidally modulated plasmonic waveguide, while Fig. 5.6b displays the measured S-parameters. The efficiency of power radiation to free space is indicated by the transmission coefficient S_{21} and reflection coefficient S_{11}, both of which are below -10 dB between 8.4 and 10 GHz. Since the plasmonic waveguide losses are very little at microwave frequencies, we can calculate the radiation efficiency from the S-parameters. The simulated radiation efficiency can be computed with a minimum of 90% from 8.4 to 10 GHz and a maximum of 95% from 9 to 10 GHz, as shown in Fig. 5.6b. It is also possible

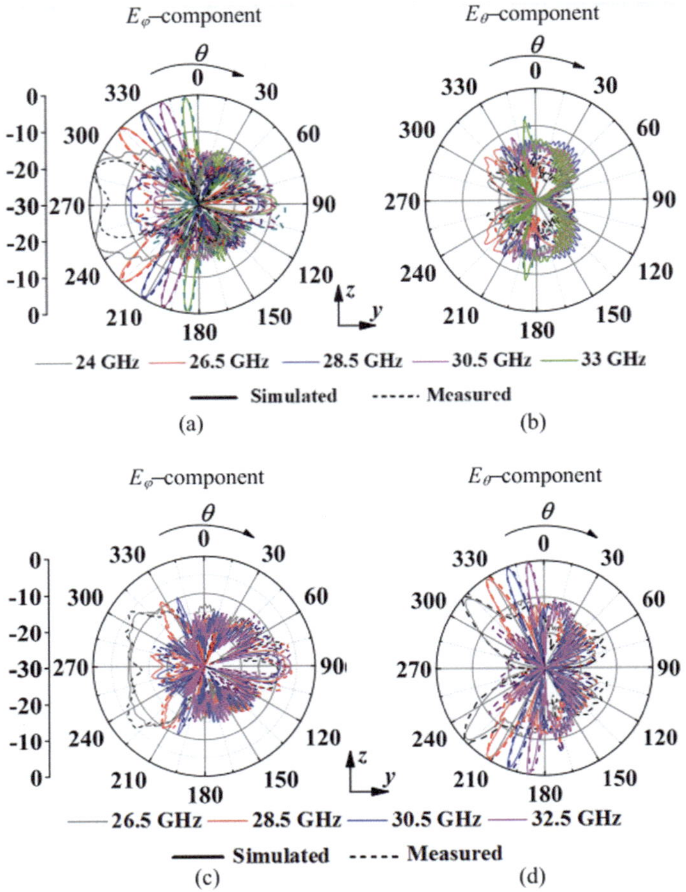

Fig. 5.4 Simulated and measured radiation pattern of SSPPs-fed open-loop [23]

to calculate the measured radiation efficiency, which is 95% from 8.8 to 10 GHz and 86% from 8.4 to 8.8 GHz. The measured far-field patterns of the E-plane in the $\upsilon = 0$ (xoz) and $\upsilon = 90°$ (yoz) planes are displayed in Fig. 5.6c–d. Figure 5.7 illustrates the use of an asymmetrical plasmonic waveguide to cancel the reflections.

According to the measurement results, the radiation efficiency is significantly higher than that of the symmetrical structure depicted in Fig. 5.8a, and both the transmission efficient (S_{21}) and reflection coefficient (S_{11}) are lower than -10 dB from 8.5 GHz to 9.8 GHz. Figure 5.8b displays the measured far-field radiation patterns of the E-plane in the yoz-plane (the plane of $\phi = 90°$). It is evident from these patterns that the main radiation lobe is directed toward 100°, 90°, and 82° at 8.5, 9.3, and 9.8 GHz, respectively. The ability of leaky-wave to continuously scan is illustrated in Fig. 5.8c. By switching the frequency from 8.5 to 9.8 GHz, the main radiation lobe can precisely steer from the backward quadrant (100° at 8.5

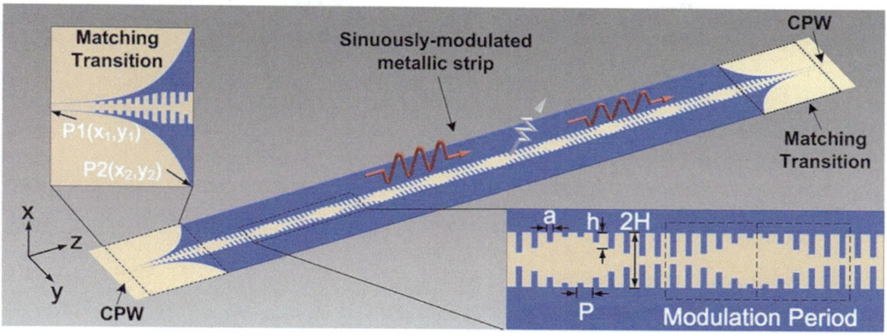

Fig. 5.5 Schematic of sinusoidally modulated plasmonic waveguide using the corrugated metallic strip [31]

GHz) to the forward quadrant (82° at 9.8 GHz) through the broadside direction (90° at 9.3 GHz), with a radiation efficiency of about 95% at broadside. The radiation angle can steer from 100° to 82°, and the measured gain is approximately 13.5 dBi, as shown in Fig. 5.8d, which displays the measured gains and radiation angles at various frequencies.

5.3.1 SSPP Unit Cell to Obtain a Wide-Angle Scanning LWA

Here, the unit cell that is used to realize LWA has a ground plane on the back. Here the unit cell confines fields as compared to that achieved with the groundless unit cells. SSPP unit cells can be of any shape and size; the effects of variation in shape and size are studied and discussed with the help of dispersion diagrams. The asymptotic cut-off frequency of the unit cell can be varied by varying the geometrical parameters. Here, we have studied three different types of corrugations. Geometries with three different corrugations namely corrugations I, II, and III with the slit width (a) of 2.25, 1.25, and 0.25 mm, respectively are shown in Fig. 5.9a. The groove depths (h) are 3.9, 2.95, and 2 mm for corrugation I, II, and III, respectively with a fixed periodicity (p) of 3 mm. Figure 5.9b shows the dispersion diagrams for three different corrugations for a cut-off frequency (f_c) of 13.4 GHz. It can be observed from Fig. 5.9b that the slope of the dispersion curve for corrugation III has a large variation in phase with a small variation in frequency which leads to a wide scanning range. Also, the fields are much more confined in the case of corrugation III compared to those obtained with other corrugations.

Asymptotic cut-off frequencies of such structures are mainly dependent on two parameters: shape and groove depth. In theory, a homogeneous anisotropic medium is defined as $\varepsilon_x = p/a$ [33], where 'p' is the periodicity and 'a' is the width of slit. When the corrugation shape changes or the width of the slit changes, the effective permittivity of the structure changes hence the asymptotic frequency changes. The

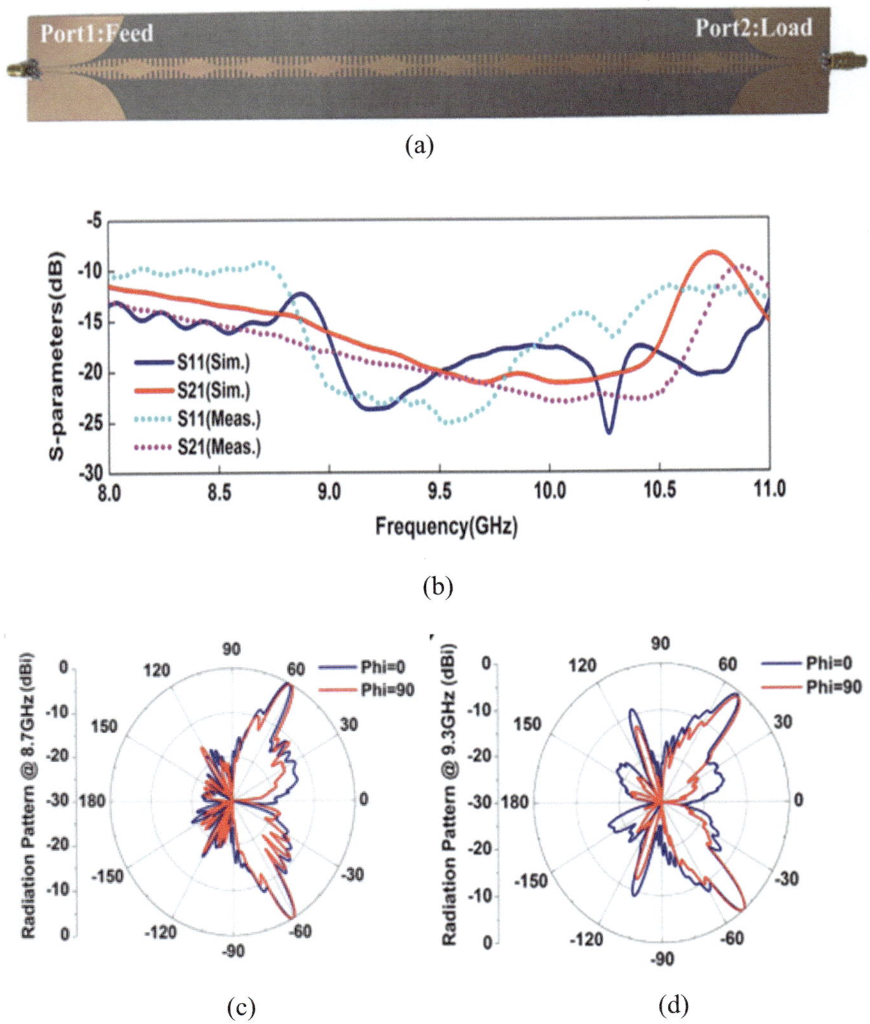

Fig. 5.6 **a** Photograph of the fabricated antenna. **b** The measured and simulated S-parameters. **c–d** The measured far-field radiation patterns of the E-plane in both $\phi = 0$ and $\phi = 90°$ planes at 8.7 and 9.3 GHz with radiation angles of 63.9° and 54.9°, respectively [31]

dispersion characteristics of different SSPP unit cells show the property of SSPP at microwave frequencies as shown in Fig. 5.10. The Eigenmode solver of CST is used to obtain the dispersion curves.

Figures 5.10a–c shows the effects of variations in groove depth for corrugation I, II, and III. It can be observed from the dispersion diagrams that the cut-off frequency increases with a decrease in the groove depth. Whereas when the width of the slit reduces, the cut-off frequency reduces hence it results in a decrement in the slope of the dispersion diagram. From the above observations, we can conclude that when the

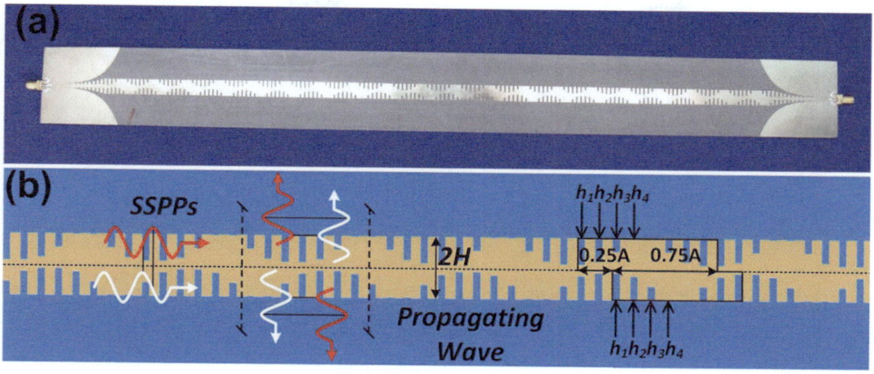

Fig. 5.7 **a** The photograph of the fabricated sample. **b** The detailed structure with different groove depths of h_1, h_2, h_3 and h_4 [31]

electrical length of the unit cell reduces the asymptotic cut-off frequency increases, and the electrical length is inversely proportional to the frequency. We conclude that corrugation I can be preferred for higher frequencies. Whereas corrugation III could be more efficient for realizing at a lower frequency compared to the other two corrugations. For getting operation at higher frequencies, the groove depth in corrugation III becomes smaller compared to that in corrugation I and II. Corrugation III will be difficult to be used at higher frequencies.

Figures 5.11a–c shows microstrip SSPP transmission lines realized by considering three different corrugations (explained above). The corrugated microstrip SSPP transmission line works as a low-pass filter whose cut-off frequency can be varied by varying the groove depth and periodicity. The overall length of each transmission line (L) is 347 mm and the width (W) is 45 mm. The length of the microstrip section (L_1) is 4 mm, the length of the transition (L_2) is 30 mm and the length of the SSPP transmission region (L_3) is 279 mm. The width of the 50 Ω microstrip line (W_s) is 4 mm. A smooth transition is realized to efficiently convert a quasi-TEM mode to an SSPP mode. A plasmon mode can be seen by plotting field strength at the metal–dielectric interface. For a plasmon mode, fields decay exponentially both in the air and metal region. Figure 5.12 shows the plot of electric field strength in different regions, the electric field strength decays exponentially in both regions. Where $z > 1.535$ mm is the air region, $1.5 \text{ mm} < z < 1.535$ mm is the metal region, $z < 1.5$ mm is the dielectric region, and z is the direction normal to the interface.

The proposed transition has a good transition occupation ratio and transmission efficiency which helps in realizing an efficient leaky-wave antenna (LWA). The transition is robust as the property of the transmission line does not change with the change in the shape of the unit cell when designed for a particular cut-off frequency. In the tapered transition region, the groove depth is varied from h_1 to h_{10}, i.e., 0.39–3.9, 0.295–2.95, and 0.2–2 mm for corrugation I, corrugation II, and corrugation III, respectively, as shown in Fig. 5.11. The performance of each transmission line in

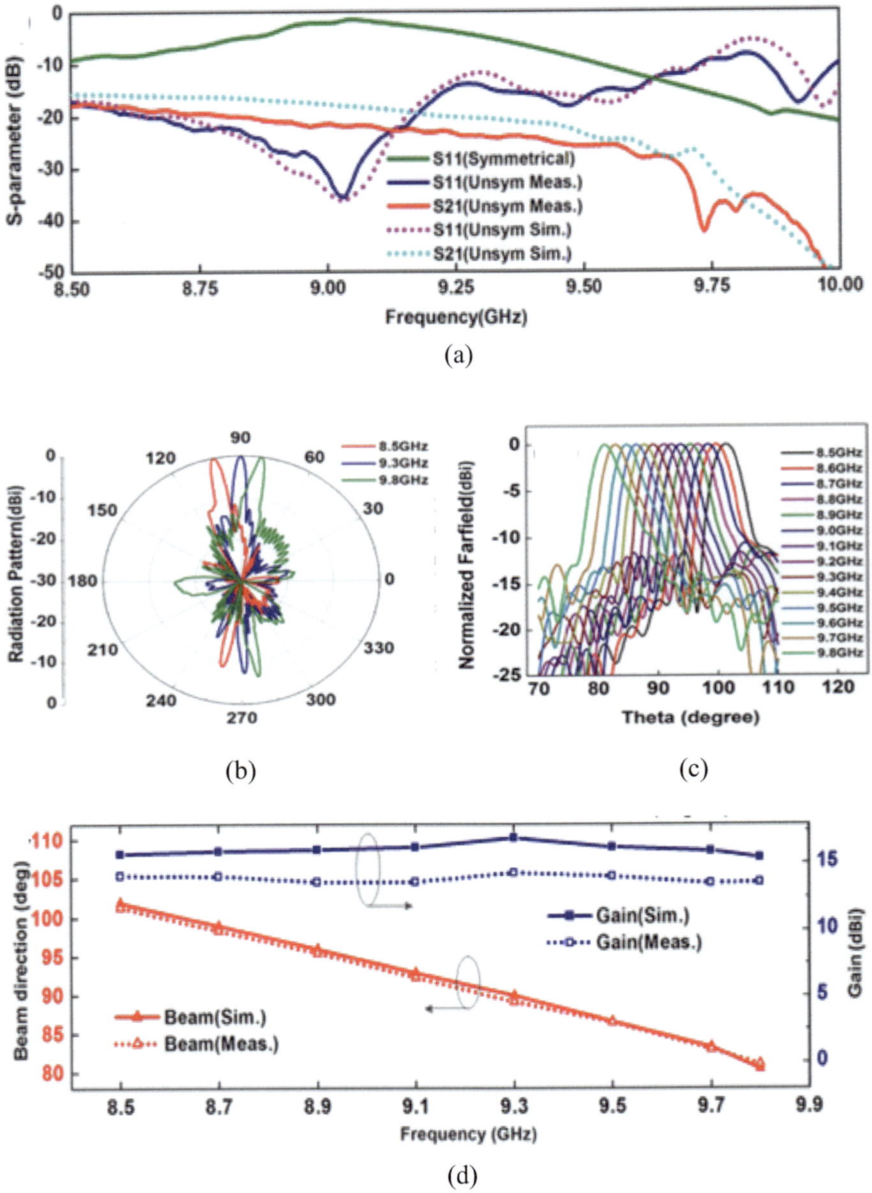

Fig. 5.8 **a** The measured S-parameters show that the reflection is reduced greatly by using an asymmetrically plasmonic waveguide. **b** The measured far-field radiation patterns of the E plane, which show that the main lobes are directed to 100°, 90°, and 82° at 8.5GHz, 9.3GHz, and 9.8GHz, respectively. **c** The main lobe can steer from backward to forward, through broadside (90° at 9.3 GHz) continuously as frequency changes from 8.5 to 9.8 GHz. **d** The measured gains and directions at different frequencies [31]

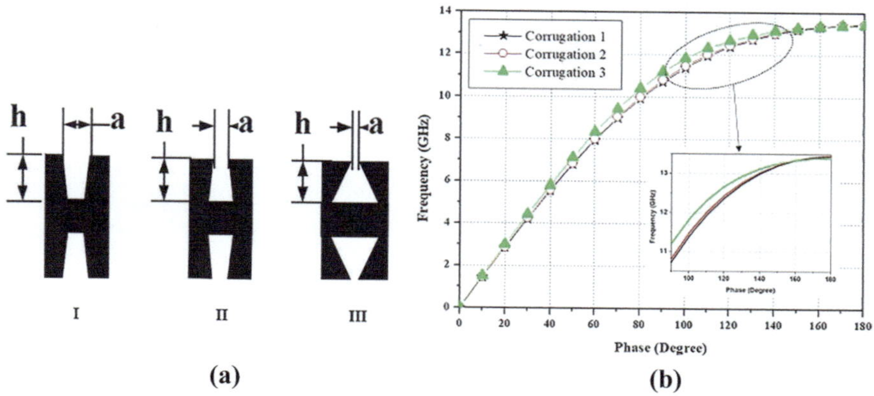

Fig. 5.9 **a** Corrugation I, II and III. **b** Dispersion diagrams for three different corrugations with a cut-off frequency of 13.4 GHz [32]

terms of reflection and transmission coefficients is shown in Fig. 5.13. Each transmission line has similar performances in terms of S-parameters. All three corrugated transmission lines are robust and efficient enough to feed the radiating elements and realize efficient LWAs.

5.3.2 *Leaky-Wave Antenna Design*

After considering different corrugated transmission lines explained in the above section, leaky-wave antennas are realized. Each antenna of size 347 mm × 45 mm is designed on a 1.5 mm thick F4B substrate with $\varepsilon_r = 2.65$ and $\tan\delta = 0.002$. Figure 5.14a–c shows the design of leaky-wave antennas with corrugation I, II, and III, respectively. The structure consists of three regions. Region I is a 50 Ω microstrip line of width 'Ws' and length 'L_1'. Region II is a transition from microstrip to SSPP TL. In this region, the groove depth is varied from h_1 to h_{10} (as shown in Fig. 5.11) to obtain efficient and smooth transition and mode conversion from quasi-TEM to SSPP mode. Region III consists of radiating patches of circular shape.

The length of the radiating portion is L_3 (279 mm). The radius of the radiating patch is r (4.8 mm) and the distance between the adjacent patches is d_1 (12 mm). The distance between the adjacent patch on the upper and lower sides is d_2 (6 mm). The distance between the patch and SSPP-TL is g (0.3 mm). The SSPP-TL has a highly confined field and it supports slow wave mode. To achieve radiation from the structure, we need to convert this slow wave mode into a fast wave mode which can be done by introducing periodic perturbation along the SSPP-TL which results in an infinite number of space harmonics given by

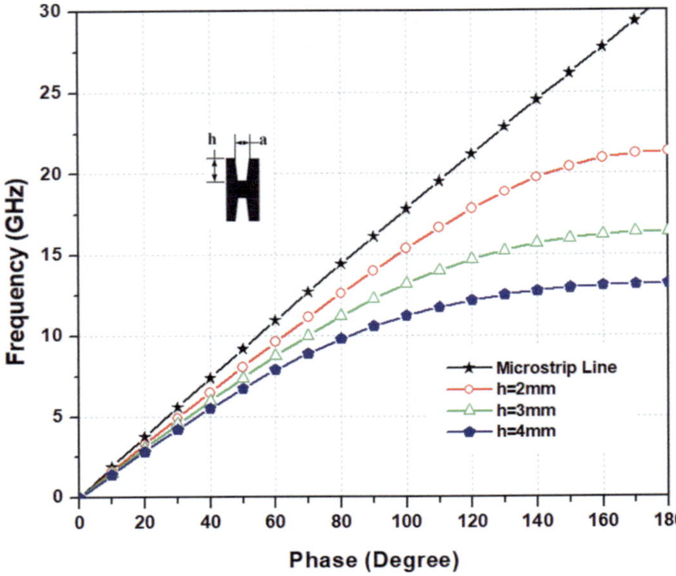

(a)

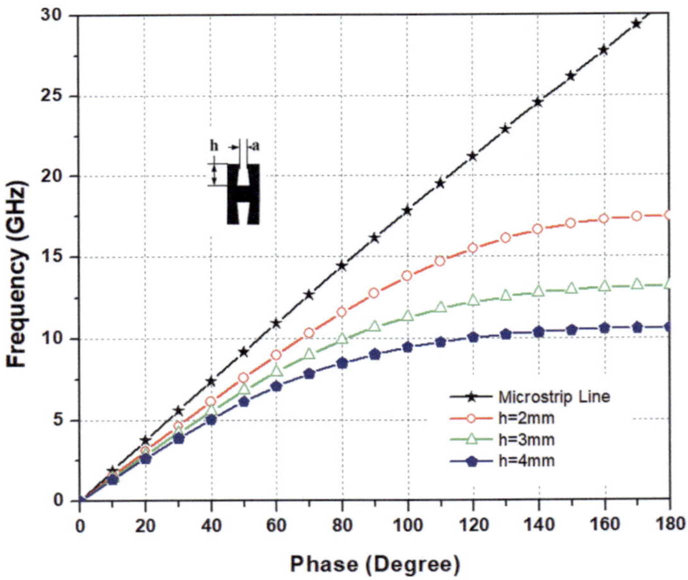

(b)

Fig. 5.10 Dispersion diagrams showing the effects of variations in groove depth for three different corrugations **a** corrugation I **b** corrugation II **c** corrugation III [32]

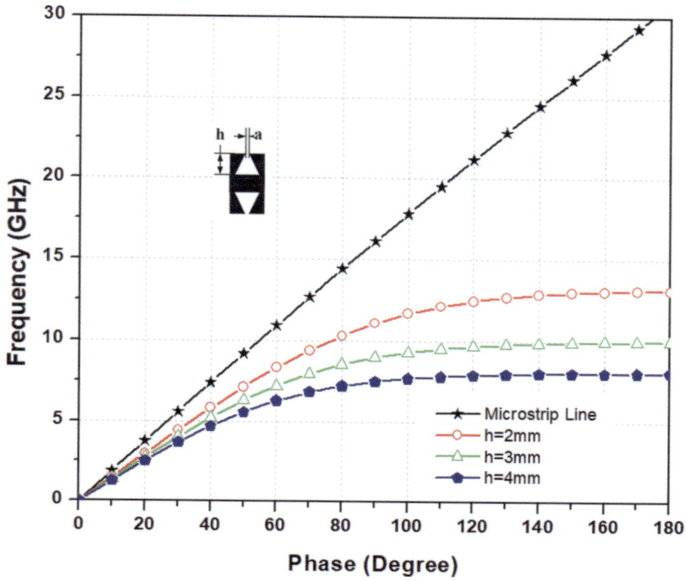

(c)

Fig. 5.10 (continued)

$$\beta_n = \beta_{SPP} - \frac{2n\pi}{p} \tag{5.1}$$

where p is the period of perturbation, β_{SPP} is the propagation constant of SSPP mode, and n is the harmonic order. In (5.1) $n = -1$ is taken for a single beam

$$\beta_{-1} = \beta_{SPP} - \frac{2\pi}{p} \tag{5.2}$$

β_{-1} can be either negative or positive depending upon p, the radiating beam can be steered in forward or backward direction. The main beam direction can be calculated from (5.3)

$$\theta = \arcsin\left(\frac{\beta_{-1}}{K_0}\right) \tag{5.3}$$

In this design, circular patches are used to convert an SSPP mode into a radiating mode. The radius of a circular patch (r) is used to determine the resonant frequency of the antenna.

The resonant frequency of the antenna can be calculated as given in [30]

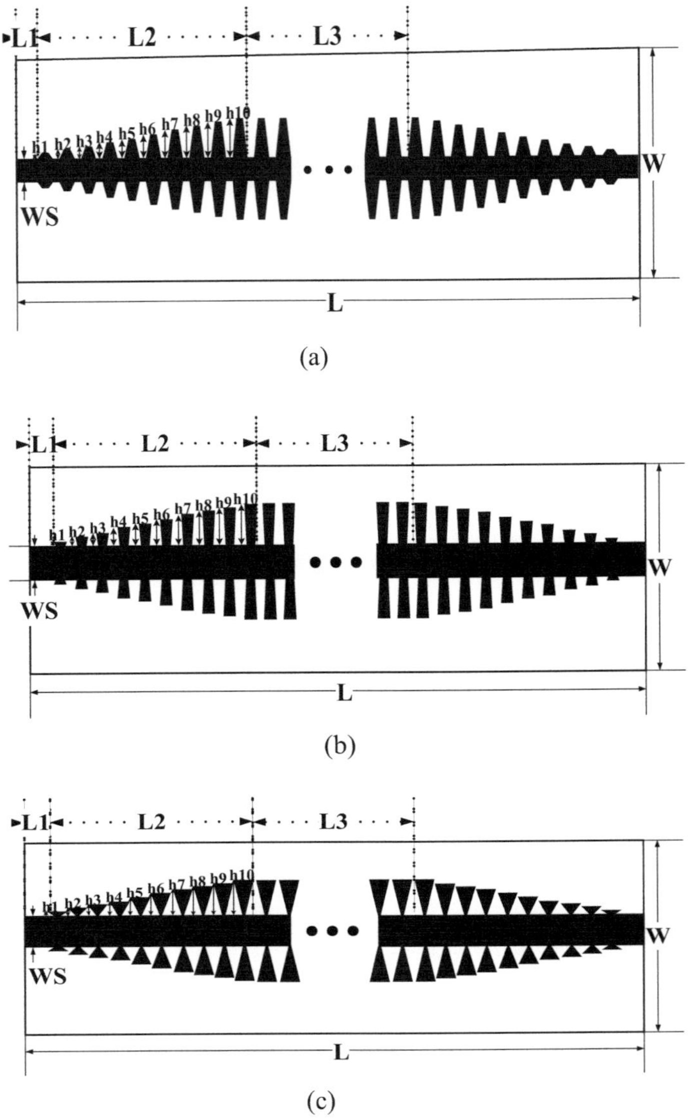

Fig. 5.11 Microstrip SSPP transmission lines were designed by considering **a** corrugation I **b** corrugation II **c** corrugation III [32]

$$f_r = \frac{c}{2\pi\sqrt{\varepsilon_r}}\left(\frac{\chi'_{\mathrm{mn}}}{r}\right) \tag{5.4}$$

where χ'_{mn} represent the order of the resonant frequency and c is the speed of light. From (5.4), for the dominant mode (TM_{11}), the value of 'r' is calculated to be 5, 5, and 4.8 mm for corrugation I, II, and III, respectively. The optimized values of 'r'

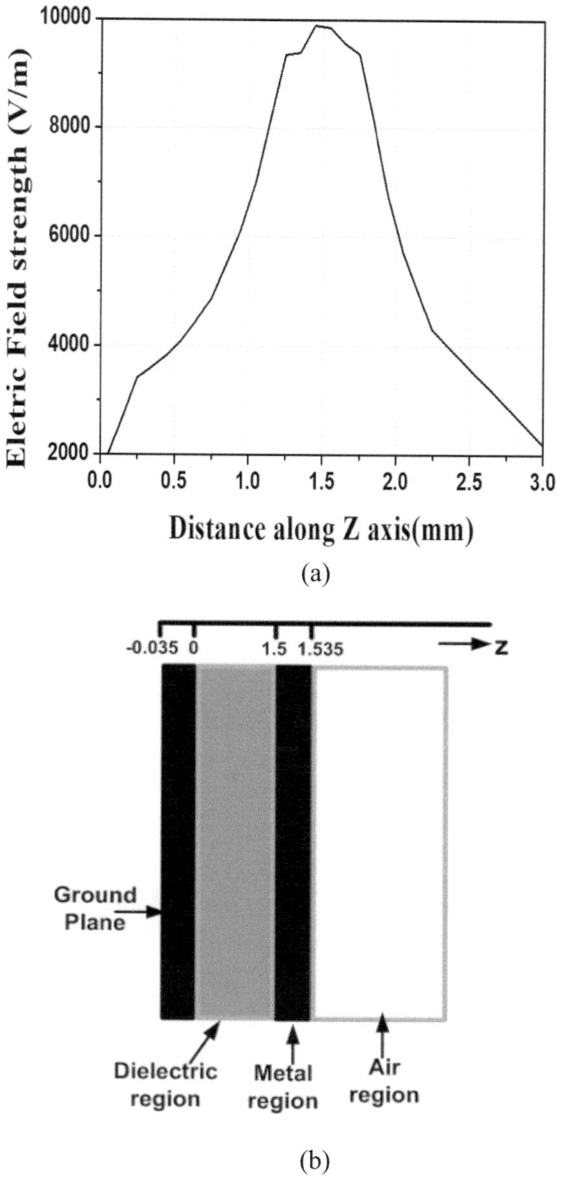

Fig. 5.12 **a** Electric field strength in various regions (air, dielectric, and metal). **b** Showing different regions [32]

for all three corrugations are 4.8 mm. The guided wavelengths (λ_g) at 10.7, 10.8, and 11.2 GHz is 12 mm for corrugation I, II, and III, respectively, (from Fig. 5.10).

The distance between the adjacent patches is kept λ_g and the distance between the patch at the upper side and the adjacent patch to the lower side is kept $\lambda_g/2$ as the phase difference of 180° is obtained between the fields at the upper side to that at the lower side [31, 34]. In this way, the reflections generated by upper and lower

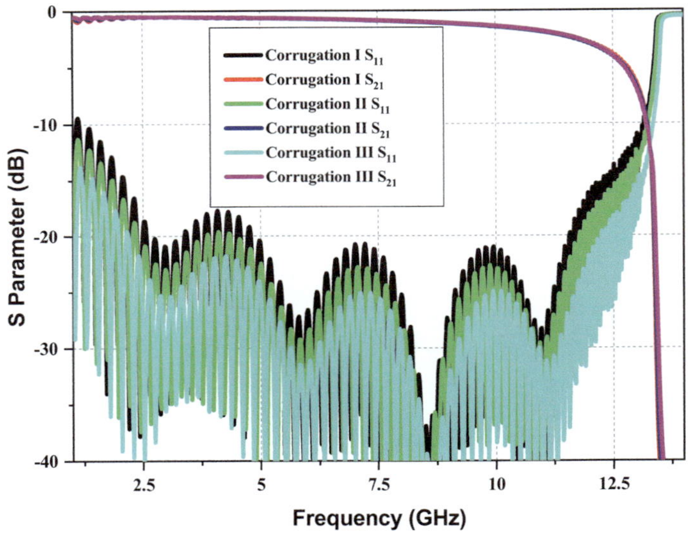

Fig. 5.13 Simulated S-parameters of transmission lines with different corrugations [32]

grooves are canceled from each other hence high radiation efficiency is achieved at the broadside. It eliminates the OSB region. The distance between the transmission line and patch is set to be 0.3 mm as the field in the case of the SSPP line is much more confined compared to the conventional transmission line. The radiating patches are excited through electromagnetic coupling.

5.4 Methodology to Achieve High-Gain Antenna Using Spoof Surface Plasmon Polariton

The design procedure for leaky-wave antenna to achieve high gain and wide scan range using many radiating elements in a limited area is given below:

a. Design an SSPP unit cell having an asymptotic cut-off frequency slightly greater than the upper limit of the desired band. For example, in this case, the asymptotic cut-off frequency is 13.4 GHz whereas the desired band is chosen from 8 to 12.5 GHz.

b. From the dispersion curve, the guide wavelength can be calculated. It is an important parameter as the distance between the patches is equal to λ_g and it can be calculated as

$$\lambda_g = (2 * \pi * p)/\beta \tag{5.5}$$

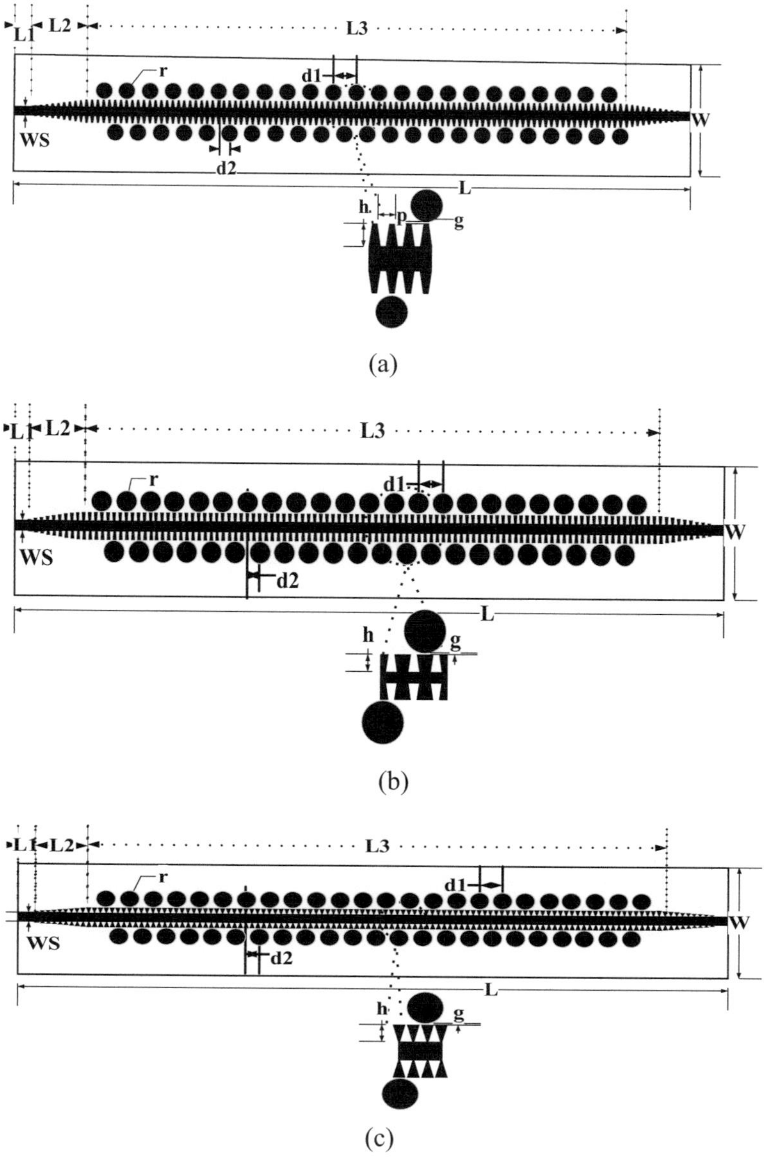

Fig. 5.14 Leaky-wave antennas with **a** corrugation I **b** corrugation II **c** corrugation III [32]

For example, here the value of periodicity (p) is 3 mm, β should be as large as possible because as we go for larger β, the slow wave characteristic can be achieved which provides a high scan range. There is also a limit for β beyond a particular value after that we cannot increase its value. Table 5.1 shows the impact of β on other parameters of the antenna. It can be observed that higher values of β are not

Table 5.1 Impact of β on antenna parameters

β (Deg)	λ_g (mm)	Cut-off frequency (fc) (GHz)	Radius (r) (mm)
180	6	13.4	4
140	7.714	13	4.156
120	9	12.377	4.365
90	12	11.229	4.8

feasible as the adjacent patches will overlap; the optimum value of β is chosen here. Figure 5.15a–c shows the zoomed view of the radiating patch's locations with respect to the MS SSPP line. Here, three different cases are considered by taking different values of β. In Fig. 5.15a, $\beta = 180°$ and $\lambda_g = 6$ mm. For this case, the calculated value of d_1 is 6 mm, d_2 is 3 mm, and the radius of the patch (r) is 4 mm, the overlapping of patches is noticed as shown in Fig. 5.15a. In the second case, $\beta = 120°$, $\lambda_g = 9$ mm so the calculated value of d_1 is 9 mm, $d_2 = 4.5$ mm, and the radius of the patch is 4.365 mm, in this case, the radiating patches are not overlapped with each other but the bottom patches are coupled with two different edges of the corrugation hence the coupled fields to the radiating patch change which results in a deteriorated radiation pattern. In the third case, $\beta = 90°$, $\lambda_g = 12$ mm, the calculated value of d_1= 12 mm and $d_2 = 6$ mm, and the radius of the patch is 4.8 mm. In this case, the coupling is symmetric for both top and bottom patches as shown in Fig. 5.7c.

This is the optimized value of β for which the maximum number of patches can be incorporated in the given area without getting any overlap or mismatch in the coupled fields. Figure 5.16 shows the simulated S-parameters of the leaky-wave antennas designed using different corrugated SSPP-TLs. All these LWAs can be used as transmission lines in the frequency range of 1–8 GHz, the insertion loss is low in this frequency range. Resonances arise around 8–10 GHz because of the coupling

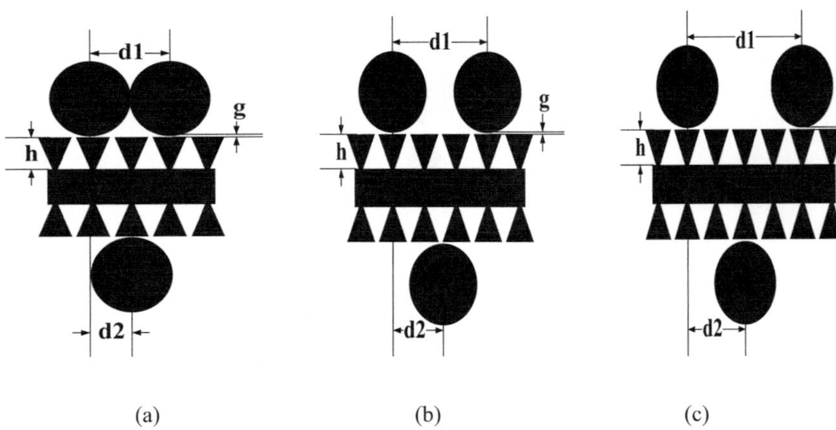

Fig. 5.15 Radiating patch locations with respect to the MS SSPP line **a** d_1= 6 mm and d_2= 3 mm **b** d_1= 9 mm and d_2= 4.5 mm **c** d_1= 12 mm and d_2= 6 mm [32]

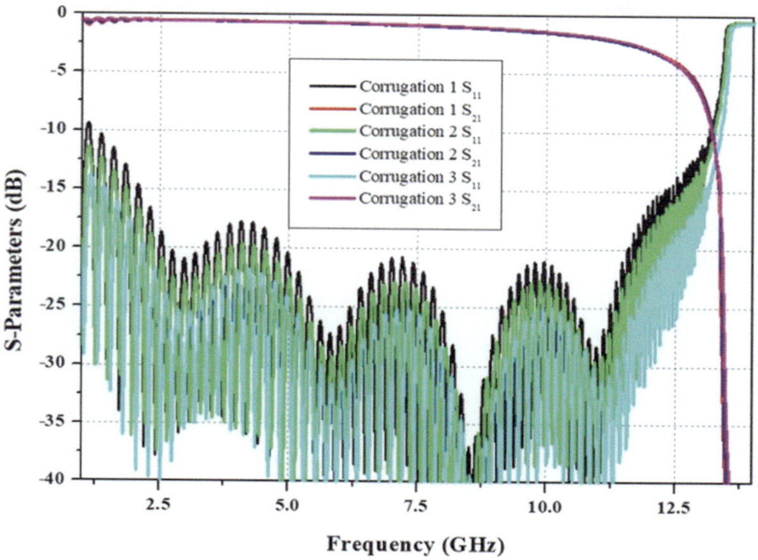

Fig. 5.16 Simulated S-parameters of LWAs with different corrugated transmission lines [32]

of fields from the SSPP-TL to the radiating patches. A radiating mode is converted from an SSPP mode by placing an array of circular patches.

These patches are electromagnetically fed through SSPP-TL. The field is coupled to the radiating patches across the operating frequency range of 8.5–12.7 GHz. Resonances are generated due to the loading effect of patches, this loading is different at different frequencies, the generation of resonance peaks also depends on the shape of the corrugation and the coupling gap between the corrugation and radiating patch. Optimized values are taken to realize a structure to be operated in a wide frequency range. Simulated realized gains of leaky-wave antennas are plotted in Fig. 5.17. Realized gain drops at those frequencies where the coupling is not good (e.g., 9.5 and 11.3 GHz for corrugation III). Antenna realized with corrugation III provides a good, realized gain in the wide frequency range of 8.25–12.7 GHz. After comparing the performance as given in Table 5.2, a leaky-wave antenna with corrugation III is demonstrated.

5.4.1 *Simulated and Measured Results*

Leaky-wave antenna with corrugation III is fabricated to validate the performance. Figure 5.18 shows the photograph of the fabricated structure. The reflection and transmission coefficients of the antenna are measured by means of a vector network analyzer (PNA-E8364C). Figure 5.19 plots the simulated and measured reflection and transmission coefficients. It is clear that the reflection coefficients are below −10

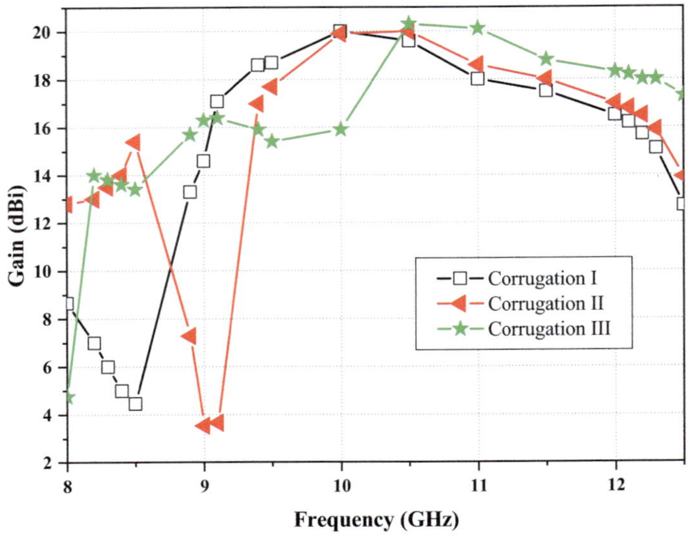

Fig. 5.17 Simulated realized gains of leaky-wave antennas [32]

Table 5.2 Comparison of the performance of leaky-wave antenna with different corrugations

Parameters	Corrugation I	Corrugation II	Corrugation III
Scan range (Deg)	66	69	92
Freq range (GHz)	8.9–12.2	9.3–12.4	8.25–12.7
Gain variation (dBi)	14–20	14–20	14–20
Efficiency	90%	90%	90%

dB in the frequency range of 1–7 GHz with low insertion loss. The structure can be used as transmission line in the range from 1 to 7 GHz. Reflection and transmission coefficients in the frequency range of 8.2–12.8 GHz show that there is no transmission of RF power to the other port of the line. This RF power is coupled to the radiating patches and it can be analyzed from the radiation patterns and radiation efficiency of the antenna in that frequency range. The simulated radiation efficiency of the antenna is more than 90% in the frequency range of 8.9–12.7 GHz. Radiation patterns of the proposed antenna are measured in an anechoic chamber after satisfying far-field conditions. The simulated and measured normalized H-plane radiation patterns are plotted in Fig. 5.20. The simulated beam is scanned from −40° to 52° while the measured beam is from −35° to 51°. A measured wide scanning range of 86° has been achieved. The slope of the dispersion curve for corrugation III tends to zero and provides a wide variation in phase for a small change in the frequency and from (5.3) it is clear that when β increases, the scan range increases. Figure 5.21a shows the comparison between the measured and simulated scan angles and simulated and measured results show a good agreement. Measured and simulated realized gains are

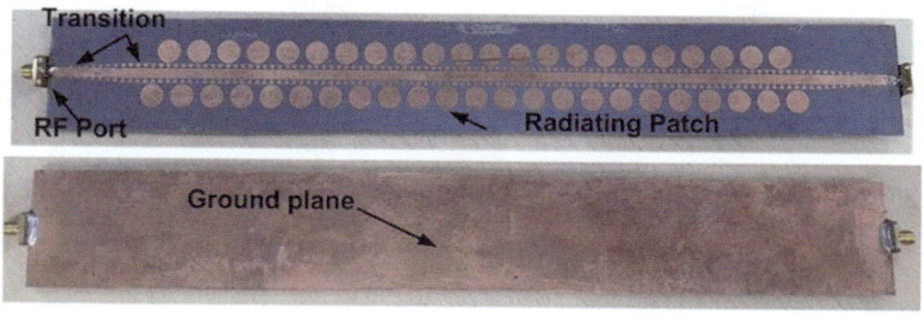

Fig. 5.18 Photograph of the fabricated leaky-wave antenna with corrugation III [32]

plotted in Fig. 5.21b. The simulated gain of the proposed antenna varies from 16 to 20 dBi while the measured gain is from 15 to 19.03 dBi in the frequency range of 8.9–12.7 GHz. The discrepancy between the simulated and measured results is mainly due to the loss added by the connector which was not considered in the simulations and fabrication imperfections.

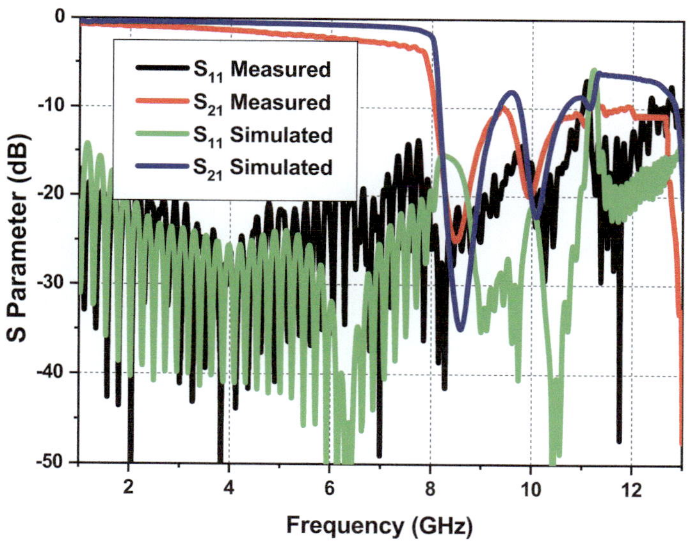

Fig. 5.19 Measured and simulated S-parameters of the proposed leaky-wave antenna with corrugation III [32]

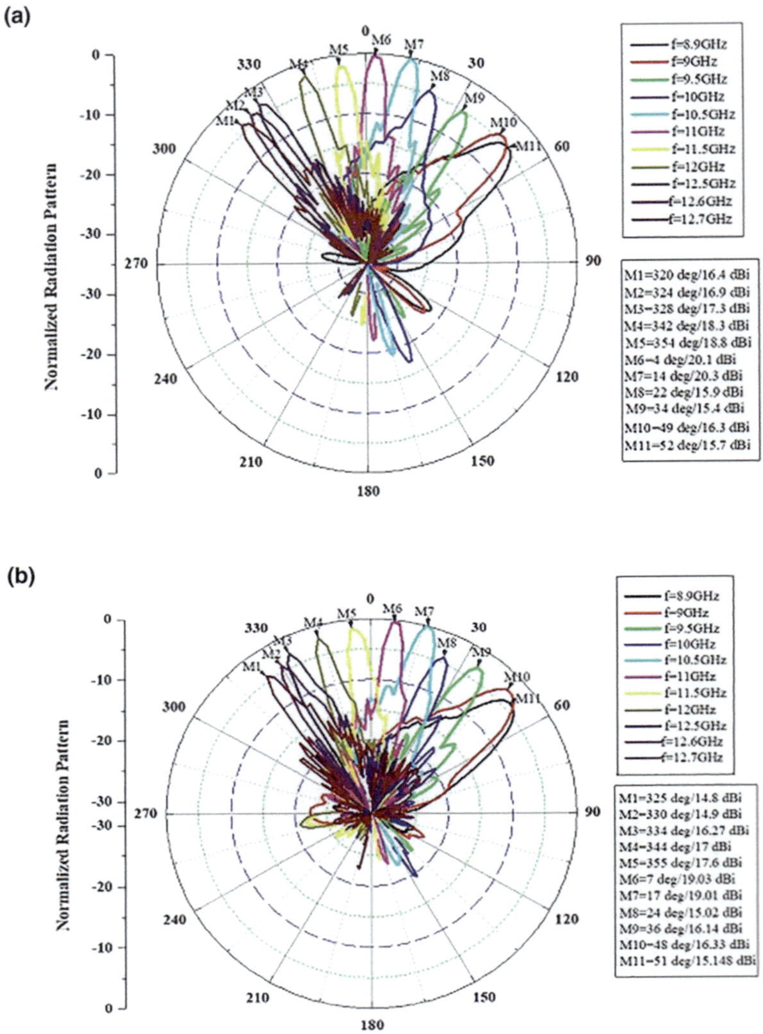

Fig. 5.20 Normalized H-plane radiation patterns of the antenna **a** simulated **b** measured [32]

5.5 Standing Wave Antenna Based on Spoof Surface Plasmon Polariton

Standing wave leaky-wave antenna (SWLWA) design provides wide bandwidth for lesser element arrays. The standing wave antenna is proposed in [35] with a beam scanning angle of 83° from 9.2 to 16 GHz with a gain of 10.3 dBi and the size of the structure is 351 mm × 59.72 mm. In this section, both standing and traveling wave antennas [36, 37] are considered for analyzing their performance in terms of

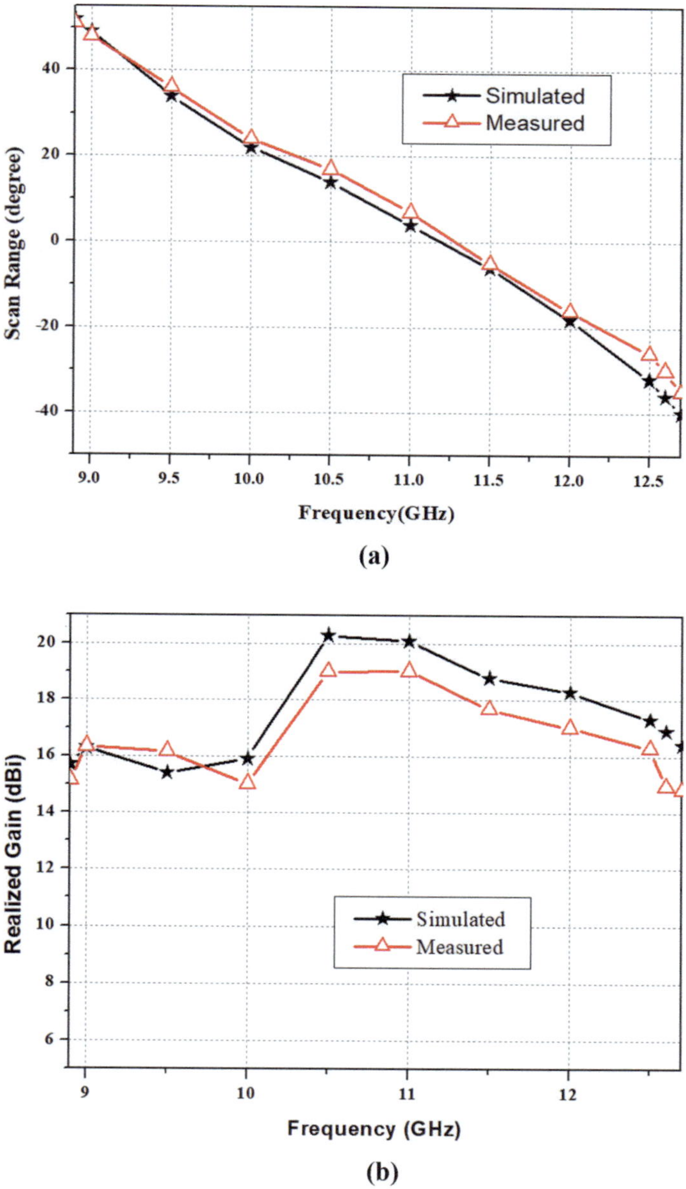

Fig. 5.21 Simulated and measured **a** scan angle **b** realized gain [32]

S-parameters, radiation pattern, scan range, and gain. Additionally, the crosstalk between two different transmission lines is studied. Microstrip, symmetrically and asymmetrically corrugated SSPP lines are taken and the performance of these lines is compared. The presented study shows that the crosstalk between the two symmetrically corrugated SSPP lines is less compared to that achieved with asymmetrically corrugated SSPP and microstrip lines. Further, these lines are used to design standing and traveling wave antennas, and again the performance of the antennas is compared. Symmetrically corrugated SSPP line-based traveling wave antenna provides a large scan range and impedance bandwidth. A standing wave antenna realized using the same symmetrically corrugated SSPP line further improves the scan range. The demonstrated standing wave antenna operates in the frequency range of 10–16 GHz with a scan range of 105°, and an average gain of 15 dBi. The advantage of the antenna is its small size compared to that reported in the literature.

5.5.1 Unit Cell and SSPP Transmission Line

Figures 5.22a, b show the geometry of an asymmetrically and symmetrically corrugated unit cell to realize an SSPP line. The geometry is designed on a 1.5 mm thick substrate with $\varepsilon_r = 2.65$ and $\tan\delta = 0.0015$. The width of the unit cell (w_s) is 4.2 mm, the height of corrugation is 'h', and the periodicity (p) of the unit cell is 2.5 mm. The width (a) of the groove is 2 mm on the upper side and 0.5 mm (b) on the lower side. The asymmetrically corrugated unit cell is corrugated at one side whereas the symmetrically corrugated unit cell is corrugated at both sides of the transmission line. The dispersion diagram of the asymmetrically and symmetrically corrugated unit cell is shown in Fig. 5.23a, b, respectively.

It is observed when the cut-off frequency of the unit cell increases, the asymptomatic cut-off frequency decreases. A highly confined field is achieved at $h = 1.6$ mm as the variation in phase is large with small variations in the cut-off frequency. The asymmetrically and symmetrically corrugated SSPP transmission lines are shown

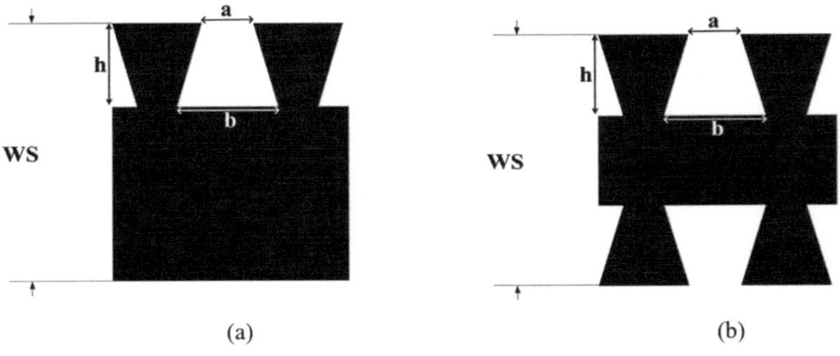

Fig. 5.22 **a** Asymmetrically corrugated unit cell and **b** Symmetrically corrugated unit cell [36]

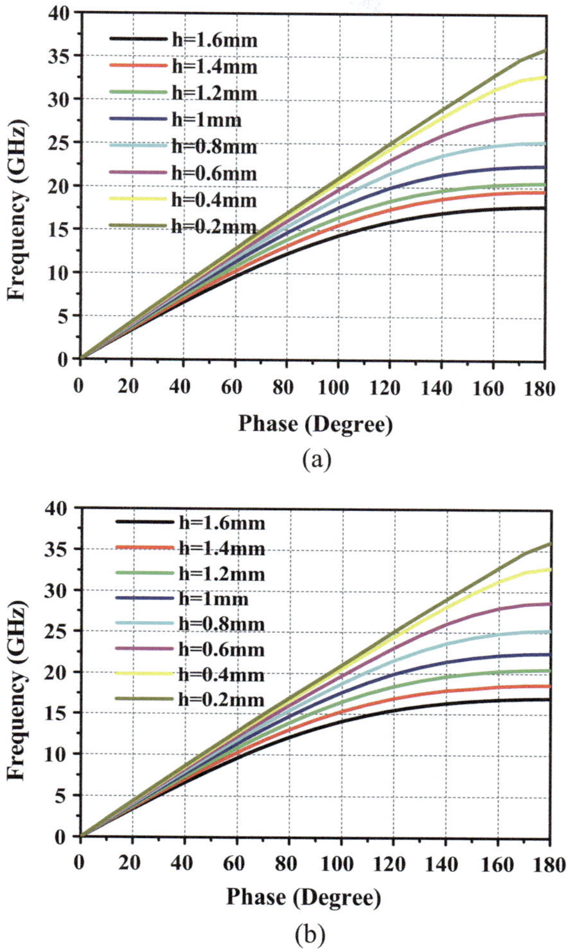

Fig. 5.23 Dispersion diagrams for **a** asymmetrically corrugated unit cell and **b** symmetrically corrugated unit cell [36]

in Fig. 5.24a, b, respectively. In the asymmetrically corrugated SSPP transmission line, corrugations are made on one side whereas in the symmetrically corrugated SSPP transmission line, corrugations are made on both sides of the line (line width is w_s). The transmission line consists of three sections namely I, II, and III. The first section (I) has a microstrip line of width $(w_s) = 4.2$ mm and length $(L_1) = 6$ mm, II is the transition which has a length $(L_2) = 20$ mm designed by gradually increasing the height of corrugation from $h_1 = 0.2$ mm to $h_{10} = 1.6$ mm and III section has a transmission line of length $(L_3) = 236$ mm.

Figure 5.25 shows the S-parameters of the asymmetrically and symmetrically corrugated SSPP transmission line having low-pass characteristics whose cut-off frequency is around 17 and 17.5 GHz, respectively, and it can also be verified from the dispersion diagrams. The transmission coefficient is better than -1 dB whereas the reflection coefficient is below -10 dB.

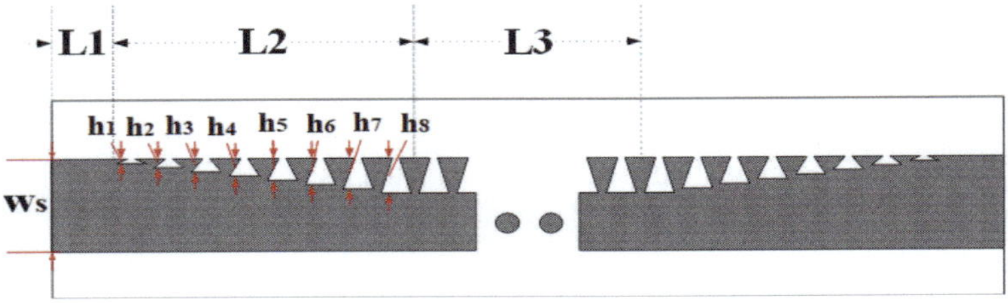

(a)

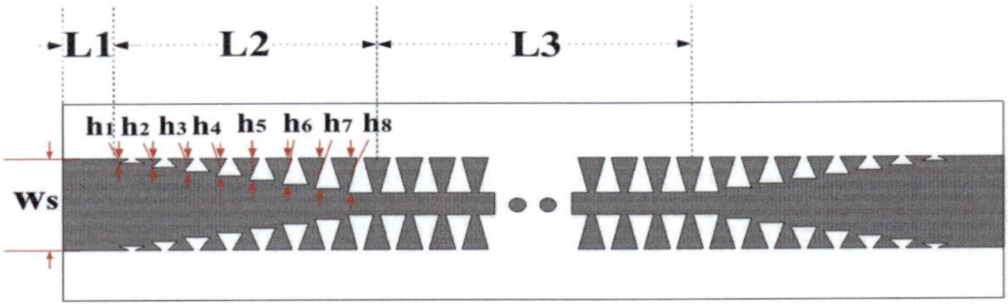

(b)

Fig. 5.24 SSPP line using **a** asymmetrically corrugated unit cell and **b** symmetrically corrugated unit cell [36]

5.5.2 *Generation of Standing Waves*

The proposed symmetrically corrugated SSPP transmission line with the change in groove size on both sides of the line at the output end to generate a standing wave is shown in Fig. 5.26. Standing wave is generated after modifying the dimensions of the unit cell. The main idea is to obtain different cut-off frequencies. The dimensions of the unit cell (in mms) are varied from a_1 to a_4 (0.5, 0.2, 0.2, 0.2) and simultaneously b_1 to b_4 (2, 3.3, 4.3, 5.3) with height (h) = 1.6 mm for a smooth transition. The highest cut-off frequency of the unit cell is around 17.8 GHz whereas the lowest cut-off frequency is 9.8 GHz which is generated by adding a structure of length (L_4) 26.5 mm. This added structure along with the transmission line is exciting a standing wave. The S-Parameters of the proposed structure are illustrated in Fig. 5.27.

5.5.3 *Antennas with Wide Scan Angle*

By considering different transmission lines explained in the above section, frequency scanning antennas are realized. Each antenna of size 298 mm × 40 mm is designed on a 1.5 mm thick F4B substrate with permittivity of 2.65 and tan δ = 0.002. Figure 5.28a–d shows the design of frequency scanning antennas using microstrip, asymmetrically corrugated SSPP, symmetrically corrugated SSPP, and symmetrically corrugated SSPP with an additional section to obtain standing waves.

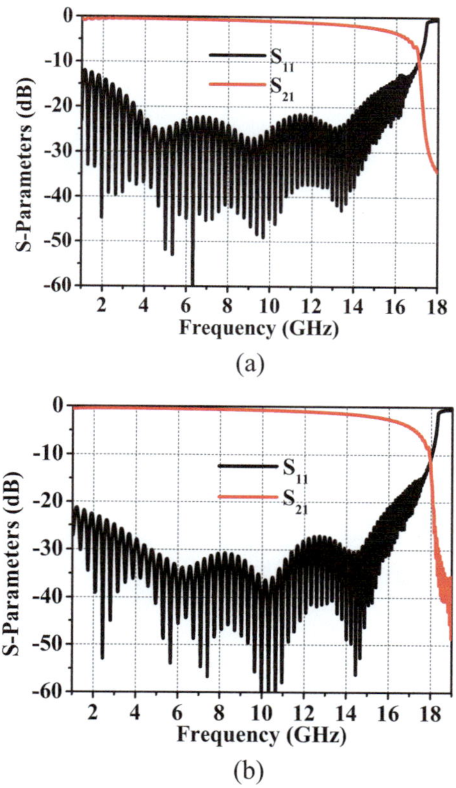

Fig. 5.25 Simulated S-parameters of the SSPP lines designed using **a** asymmetrically corrugated unit cell and **b** symmetrically corrugated unit cell [36]

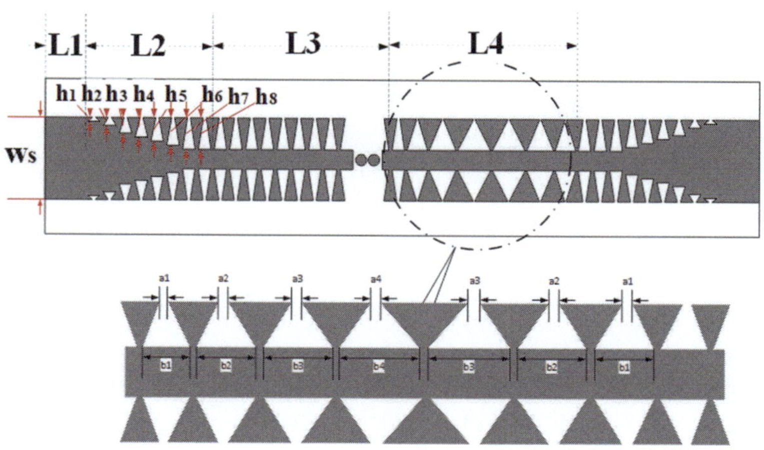

Fig. 5.26 Symmetrically corrugated SSPP transmission line with little geometrical variation for exciting a standing wave [36]

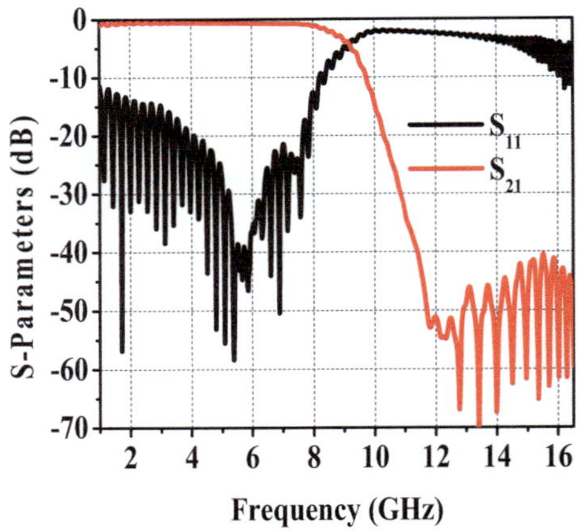

Fig. 5.27 Simulated S-parameter of symmetrical SSPP Transmission line with geometrical variation to generate standing wave [36]

The antenna is designed by loading radiating patches on both sides of the transmission line. The distance between the adjacent patches is kept λ_g (d_1) and the distance between the patch at the upper side and the adjacent patch to the lower side is kept $\lambda_g/2$ (d_2) as the phase difference of 180° is obtained between the fields at the upper side to that at the lower side. In this way, the reflections generated using upper and lower grooves are canceled from each other hence high radiation efficiency is achieved at the broadside. It eliminates the OSB region. The radius of the patch, 'r' = 3.4 mm, the value of, 'd_1' = 10 mm whereas the value of 'd_2' = 5 mm. The gap between the patch and transmission line, 'g' is 0.25 mm.

Figure 5.29 shows the simulated S-parameters of the frequency scanning antennas (FSAs) designed using microstrip line, asymmetrically corrugated SSPP line, symmetrically corrugated SSPP line, and symmetrically corrugated SSPP line with additional section. Frequency scanning antennas shown in Fig. 5.20 can be used as SSPP-TLs in the frequency range of 1–9 GHz because the insertion loss is low and return loss is better than 10 dB over the entire frequency range of 1–9 GHz. Above 9 GHz, the fields start getting coupled from an SSPP-TL to the radiating patches. A radiating mode is converted from an SSPP mode after loading a line with an array of circular patches. The RF power to these patches is electromagnetically coupled. From the S-parameters, it is clear that the coupling bandwidth is larger for symmetrically corrugated SSPP-based antenna and symmetrically corrugated SSPP-based antenna with an additional section.

The normalized radiation patterns of designed antennas are plotted in Fig. 5.30. Figure 5.30d shows that the scan range is maximum in the case of a symmetrically corrugated SSPP-based antenna with an additional section. The number of radiating elements is taken the same (40 radiating patches are used) for all designs for comparison purposes. For 3 dB gain variation, the scan angle range of microstrip, asymmetrically corrugated SSPP, symmetrically corrugated SSPP, and symmetrically corrugated SSPP with additional section-based FSAs is 50°, 85°, 95°, and 105°, respectively.

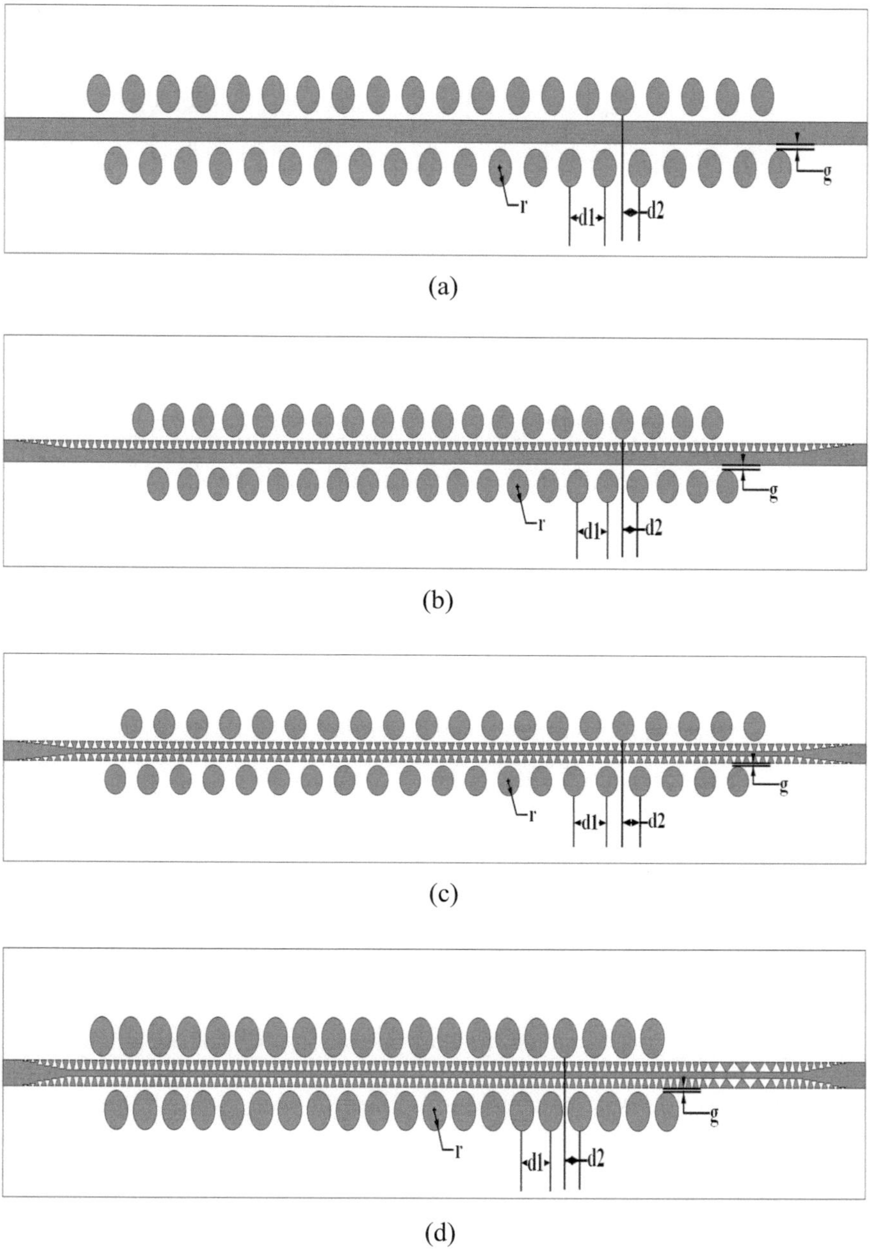

Fig. 5.28 Geometry of frequency scanning antennas using **a** microstrip line **b** asymmetrically corrugated SSPP line **c** symmetrically corrugated SSPP line **d** symmetrically corrugated SSPP line with additional section [36]

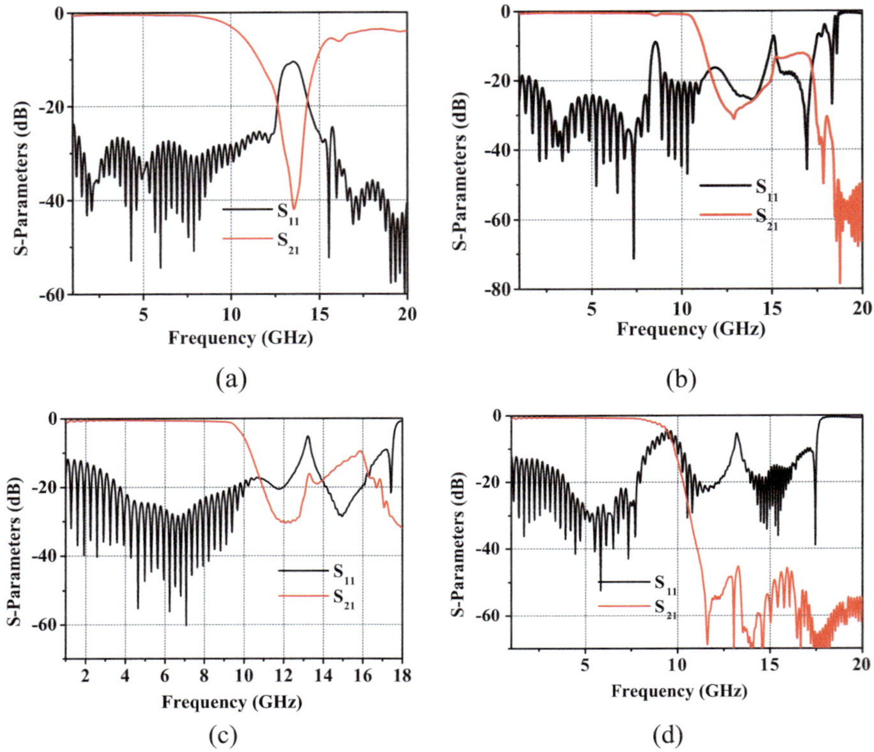

Fig. 5.29 Simulated S-Parameter of frequency scanning antenna using **a** microstrip line **b** asymmetrically corrugated SSPP line **c** symmetrically corrugated SSPP line **d** symmetrically corrugated SSPP line with additional section [36]

5.5.4 Simulated and Measured Results

Frequency scanning antennas using symmetrically corrugated SSPP with an additional section for exciting standing waves are fabricated to validate the performance. Figure 5.31 shows the photograph of the fabricated prototype. The reflection and transmission coefficients of the antenna are measured by means of a vector network analyser (PNA-E8364C). Figure 5.32 plots the simulated and measured reflection and transmission coefficients. The reflection coefficients are below −10 dB in the frequency range of 10–16.2 GHz. The RF power is coupled to the radiating patches in the frequency range of 10–16.2 GHz. The normalized radiation patterns of the proposed antenna are measured in an anechoic chamber by satisfying far-field conditions. The measured normalized radiation patterns are plotted in Fig. 5.33.

The gain and scan range comparison for all four cases are shown in Fig. 5.34a, b, respectively. Figure 5.34a shows that the gain variation is maximum for microstrip whereas minimum for symmetrically corrugated exciting standing wave frequency scanning antenna. Similarly, Fig. 5.34b shows that the scan range is minimum for

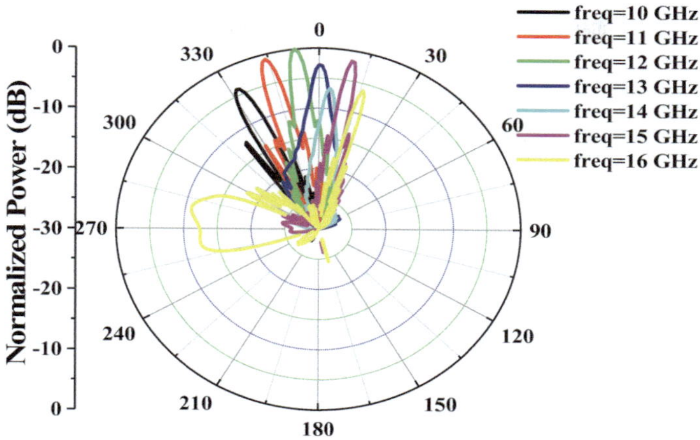

(a)

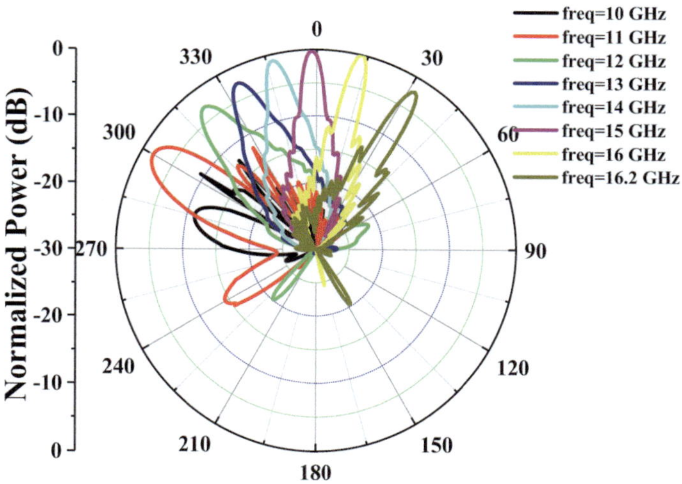

(b)

Fig. 5.30 Normalized radiation patterns of the frequency scanning antenna using **a** microstrip line **b** asymmetrically corrugated SSPP line **c** symmetrically corrugated SSPP line **d** symmetrically corrugated SSPP line with additional section [36]

microstrip whereas maximum for symmetrically corrugated exciting standing wave frequency scanning antenna.

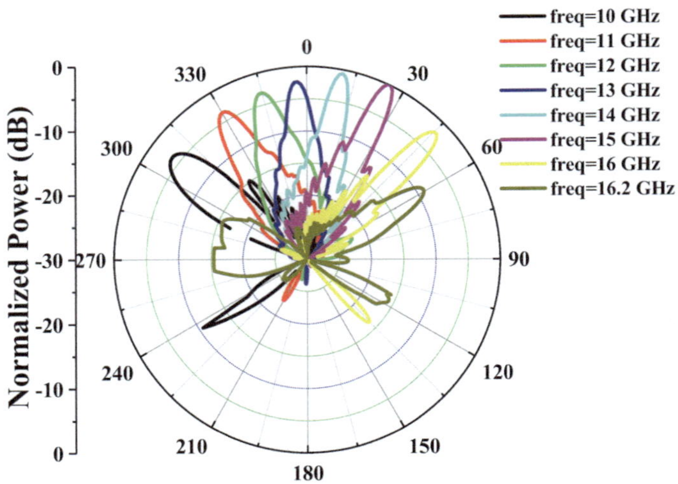

(c)

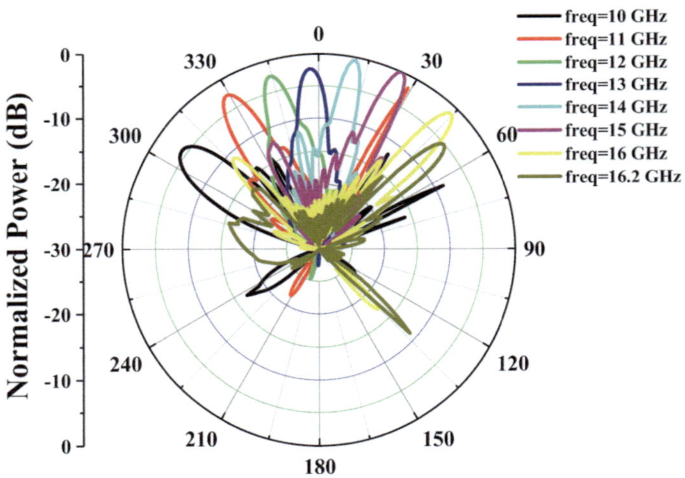

(d)

Fig. 5.30 (continued)

5.6 Conclusion

In this chapter, we have reported different types of frequency scanning antennas and methods to resolve the open stop band effects. The difference between a traveling wave antenna and a standing wave antenna is studied. The design methodology for realizing compact high-gain leaky-wave antenna is explained. The designs of

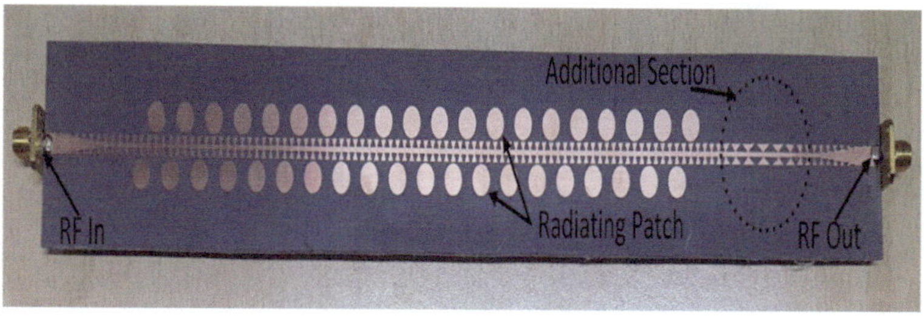

Fig. 5.31 Fabricated prototype of frequency scanning antenna using symmetrically corrugated SSPP generating standing waves [36]

Fig. 5.32 Simulated and measured S-parameters of the frequency scanning antenna exciting standing waves [36]

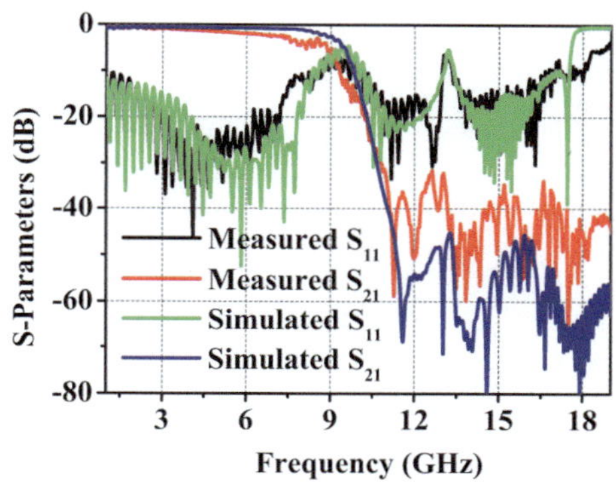

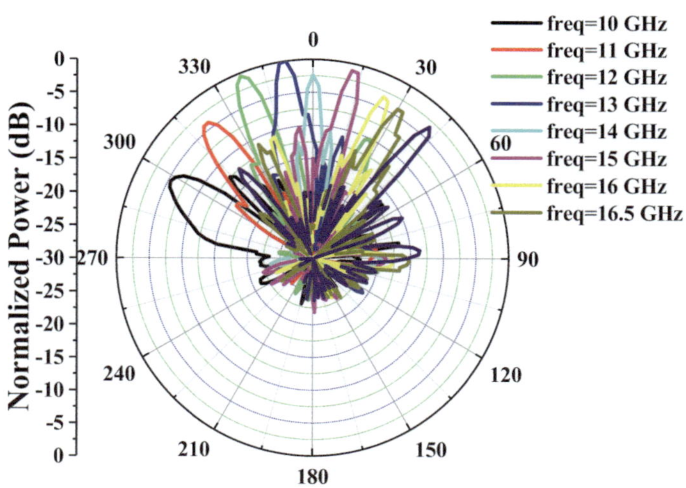

Fig. 5.33 Measured normalized radiation patterns of the frequency scanning antenna as illustrated in Fig. 5.27 [36]

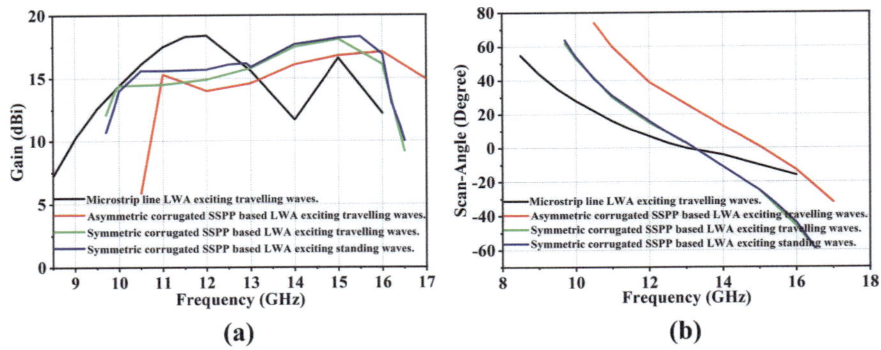

Fig. 5.34 Simulated results of the frequency scanning antenna exciting standing waves **a** realized gain **b** scan angle [36]

both traveling and standing wave antennas are explained in detail with experimental verification.

References

1. Johnson, R.C., Jasik, H.: Antenna Engineering Handbook. Mc Graw-Hill, New York, NY, USA (1993)
2. Jackson, D.R., Oliner, A.A.: Leaky-wave antennas. In: Balanis, C.A.(eds.) Modern Antenna Handbook. Wiley, Hoboken, NJ, USA (2008)
3. Baccarelli, Burghignoli, P., Comite, D., Fuscaldo, W., Galli, A.: Open-stopband suppression via double asymmetric discontinuities in 1-D periodic 2-D leaky-wave structures. IEEE Antennas Wirel. Propag. Lett. **18**(10), 2066–2070 (2019)
4. Paulotto, Baccarelli, P., Frezza, F., Jackson, D.R.: A novel technique for open-stopband suppression in 1-D periodic printed leaky wave antennas. IEEE Trans. Antennas Propag. **57**(7), 1894–1906 (2009)
5. Liu, J., Zhou, W., Long, Y.: A simple technique for open-stopband suppression in periodic leaky-wave antennas using two nonidentical elements per unit cell. IEEE Trans. Antennas Propag. **66**(6), 2741–2751 (2018)
6. Williams, J.T., Baccarelli, P., Paulotto, S., Jackson, D.R.: 1-D combline leaky-wave antenna with the open-stopband suppressed: design considerations and comparisons with measurements. IEEE Trans. Antennas Propag. **61**(9), 4484–4492 (2013)
7. Otto, S., Al-Bassam, A., Rennings, A., Solbach, K., Caloz, C.: Radiation efficiency of longitudinally symmetric and asymmetric periodic leaky-wave antennas. IEEE Antennas Wirel. Propag. Lett. **11**, 612–615 (2012)
8. Rahmani, M.H., Deslandes, D.: Backward to forward scanning periodic leaky-wave antenna with wide scanning range. IEEE Trans. Antennas Propag. **64**(7), 3326–3335 (2017)
9. Lyu, Y., Meng, F., Yang, G., Emi, D., Wu, Q., Wu, K.: Periodic SIW Leaky-wave antenna with large circular polarized beam scanning range. IEEE Antennas Wirel. Propag. Lett. **16**, 2493–2496 (2017)
10. Lyu, Y., et al.: Leaky-wave antenna based on non cut off substrate integrated waveguide supporting beam scanning from backward to forward. IEEE Trans. Antennas Propag. **64**(6), 2155–2164 (2016)

11. Yan, N., Ma, K., Zhang, H., Jian, Z.: A novel substrate integrated suspended line wideband leaky-wave antenna. IEEE Antennas Wirel. Propag. Lett. **16**, 2642–2645 (2017)
12. Xu, S., et al.: A wide-angle narrowband leaky-wave antenna based on substrate integrated waveguide-spoof surface plasmon polariton structure. IEEE Antennas Wirel. Propag. Lett. **18**(7), 1386–1389 (2019)
13. Rezaee, S., Memarian, M.: Analytical study of open-stopband suppression in leaky-wave antennas. IEEE Antennas Wirel. Propag. Lett. **19**(2), 363–367 (2020)
14. Kianinejad, A., Chen, Z.N., Qiu, C.: Low-loss spoof surface plasmon slow-wave transmission lines with compact transition and high isolation. IEEE Trans. Microw. Theory Tech. **64**(10), 3078–3086 (2016)
15. Kianinejad, A., Chen, Z.N., Qiu, C.: A single-layered spoof-plasmon-mode leaky wave antenna with consistent gain. IEEE Trans. Antennas Propag. **65**(2), 681–687 (2017)
16. Sarkar, A., Lim, S.: Annular surface plasmon polariton-based frequency-scanning leaky-wave antenna for full azimuth coverage. IEEE Trans. Antennas Propag. **70**(1), 180–188 (2022). https://doi.org/10.1109/TAP.2021.3098598
17. Zu, H.-R., Wu, B., Su, T.: Beam manipulation of antenna with large frequency-scanning angle based on field confinement of spoof surface plasmon polaritons. IEEE Trans. Antennas Propag. **70**(4), 3022–3027 (2022). https://doi.org/10.1109/TAP.2021.3118739
18. Li, Q., Yin, Y., Chen, X., Xu, K.-D.: A frequency-controlled wide-angle scanning antenna based on spoof surface plasmon polariton. IEEE Photon. Technol. Lett. **34**(23), 1300–1303 (2022) https://doi.org/10.1109/LPT.2022.3213643
19. Ye, L., Wang, Z., Zhuo, J., Han, F., Li, W., Liu, Q.H.: A back-fire to forward wide-angle beam steering leaky-wave antenna based on SSPPs. IEEE Trans. Antennas Propag. **70**(5), 3237–3247 (2022). https://doi.org/10.1109/TAP.2021.3137241
20. Liao, D., Zhang, Y., Wang, H.: Wide-angle frequency-controlled beam-scanning antenna fed by standing wave based on the cutoff characteristics of spoof surface plasmon polariton. IEEE Antennas Wirel. Propag. Lett. **17**, 1238–1241 (2018)
21. Yin, J.Y., et al.: Frequency-controlled broad-angle beam scanning of patch array fed by spoof surface plasmon polariton. IEEE Trans. Antennas Propag. **64**(12), 5181–5189 (2016)
22. Du, X., Ren, J., Li, H., Zhang, C., Liu, Y., Yin, Y.: Design of a leaky-wave antenna featuring beam-scanning from backfire utilizing odd-mode spoof surface plasmon polaritons. IEEE Trans. Antennas Propag. (2021)
23. Cao, D., Li, Y., Wang, J.: Spoof surface plasmon polaritons fed frequency-scanning open-loop antenna arrays. IEEE Access **7**, 179954–179960 (2019)
24. Zhang, Q.L., et al.: High-efficiency circularly polarised leaky-wave antennas fed by spoof surface plasmon polariton. IET Microw. Antennas Propag. **12**, 1639–1644 (2018)
25. Liu, L., et al.: Single-beam leaky-wave antenna with lateral continuous scanning functionality based on spoof surface plasmon polariton transmission line. IEEE Access **7**, 25225–25230 (2019)
26. Zhang, Q.L., et al.: Spoof surface plasmon polariton leaky-wave antennas using periodically loaded patches above PEC and AMC ground planes. IEEE Antennas Wirel. Propag. Lett. **16**, 3014–3017 (2017)
27. Wei, D., et al.: Wide-scanning-angle leaky-wave array antenna based on microstrip SSPPs-TL. IEEE Antennas Wirel. Propag. Lett. **17**, 1566–1569 (2018)
28. Guan, D., You, P., Zhang, Q., Lu, Z., Yong, S., Xiao, K.: A wide-angle and circularly polarized beam-scanning antenna based on microstrip spoof surface plasmon polariton transmission line. IEEE Antennas Wirel. Propag. Lett. **16**, 2538–2541 (2017)
29. Yu, H.-W., Jiao, Y.-C., Zhang, C., Weng, Z.-B.: Dual-linearly polarized leaky-wave patch array with low cross-polarization levels using symmetrical spoof surface plasmon polariton lines. IEEE Trans. Antennas Propag. **69**(3), 1781–1786 (2021)
30. Jackson, D.R., et al.: The fundamental physics of directive beaming at microwave and optical frequencies and the role of leaky waves. Proc. IEEE **99**(10), 1780–1805 (2011)
31. Kong, G., Ma, H., Cai, B., et al.: Continuous leaky-wave scanning using periodically modulated spoof plasmonic waveguide. Sci. Rep. **6**, 29600 (2016)

32. Sharma, S., et al.: A wide-angle forward to backward scanning wideband leaky-wave antenna based on microstrip spoof surface plasmon polaritons. IET Microw. Antennas Propag. **16**(2–3), 91–102 (2022)
33. Balanis, C.A.: Antenna theory: analysis and design, 2nd edn. John Wiley and Sons, Inc.
34. Bai, Y., Cheng, A.: A spoof surface plasmon leaky-wave antenna with circular polarization. Int. J. RF Microw. Comput. Aided Eng. (2020)
35. Liao, D., Zhang, Y., Wang, H.: Wide-angle frequency-controlled beam-scanning antenna fed by standing wave based on the cutoff characteristics of spoof surface plasmon polaritons. IEEE Antennas Wirel. Propag. Lett. **17**(7), 1238–1241 (2018)
36. Sharma, S., Singh, R.K., Basu, A., Koul, S.K.: Spoof surface plasmon polariton based frequency scanning antennas. IETE J. Res. **69**(11), 7787–7795 (2023)
37. Sharma, S., Singh, R.K., Basu, A., Koul, S.K.: Microstrip spoof surface plasmon polariton based leaky wave antenna. In: 2021 IEEE Indian Conference on Antennas and Propagation (InCAP), Jaipur, Rajasthan, India, India, pp. 897–900 (2021)

Chapter 6
End-Fire Antennas Using Spoof Surface Plasmon Polariton

6.1 Introduction

The antenna according to the radiation characteristic can be characterized basically into two types: leaky wave antennas and end-fire antennas. Leaky wave antennas are discussed in previous chapters where we can notice that SSPP mode is converted to the radiating mode by using patch arrays or periodically modulated SSPP TLs. Over the past few years, SSPP end-fire antennas have been extensively studied. In comparison to conventional end-fire antennas, these antennas offer a high degree of integration and are therefore appealing for planar-integrated applications. One popular method for achieving high-performance end-fire radiation is the use of surface-wave antennas. The dielectric rod antenna is a very typical structure among surface-wave antennas. In some applications, this rod's weight and price could be a problem. The Yagi-Uda antenna is another popular kind of surface-wave antenna. In relatively low-frequency applications, such as UHF television antennas, the Yagi-Uda antenna is highly popular due to its low cost and simple, lightweight structure.

End-fire radiation-resistant antennas are widely used in both military and commercial applications, such as radio astronomy, wireless communication, and radar. Due to their capacity to gather the target's complete polarization information across a broad frequency range, wideband dual-polarized end-fire antennas have grown in popularity along with the development of passive detection and electronic warfare systems. Many techniques have been used to design dual-linearly polarized antennas with end-fire radiation. One can attain dual-polarized radiation by utilizing components that can accommodate two orthogonal modes, such as patch, open-cavity dipoles, cross dipoles, or combining two antennas. Yagi-Uda antennas with circular polarization can also be converted to dual-polarization radiation.

Radar systems and wireless communications are just two of the many applications that make extensive use of planar end-fire antennas. Because of its high gain, small profile, and simplicity of integration with other planar circuits on a single substrate, this kind of antenna is widely used. Vertically polarized end-fire antennas

S. K. Koul and S. Sharma, *Microwave Integrated Circuits Based on Spoof Surface Plasmon Polariton*, Signals and Communication Technology,
https://doi.org/10.1007/978-981-96-9041-1_6

are preferred over horizontally polarized ones in many situations because as it move across the ground, radiation that is vertically polarized experiences less attenuation. However, the high profile, limited bandwidth, and manufacturing complexity of standard vertically polarized end-fire antennas are drawbacks. However, it will be challenging to manufacture these antennas for applications that call for higher gain at higher frequencies, as their structure will either be too complex or incompatible with standard low-cost processes, such as printed circuit board (PCB). Recent years have seen the emergence of the idea of 'spoof surface plasmon polariton' (SSPP) at relatively high frequencies. So, in this chapter, we will study in detail the design of end-fire antenna using SSPP.

6.2 End-Fire Antenna

The end-fire antenna based on SSPP with detailed design techniques is explained in this section. The advantage of using SSPP for end-fire antenna is explained with the help of different design techniques. The end-fire SSPP antenna's small size, simplicity in fabrication, and high performance make it a valuable component of SPP communication systems and integrated circuits. The characteristics of low transmission loss and subwavelength confinement have drawn a lot of interest in planar spoof surface plasmonic polaritons. In [1] the SSPP waveguide which acts as a balun in this case is used to feed the printed dipole antenna used as a main radiator. An I-shaped resonator array is loaded with the printed dipole antenna to realize a highly effective refractive index. Metamaterial loaded structure helps us to improve the gain we can achieve a gain of 7 dBi at 6 GHz in the end-fire direction.

Another method to design an end-fire antenna is an equivalent radiating aperture created by progressively narrowing the grooves of an odd-mode plasmonic structure in order to convert the bounded spoof SPPs mode to radiation mode [2]. Using a horn antenna as an example, the mechanism of the SPP wave radiating out through the aperture is explained. The antenna performs well at the mm-wave frequency, with a gain of about 15 dB and a radiating efficiency of about 90%. The end-fire direction's narrow beamwidth suggests that switched-beam arrays for 5G applications can be built with it. The SSPP structure combined with the back feeding technique results in a bidirectional end-fire antenna with 8-shaped radiation characteristic [3]. Placing parasitic strips at the ends of two dipole antennas yields the end-fire radiation. A 7 dBi realized gain and a total radiation efficiency of over 85% from 5.3 to 5.7 GHz can be observed with this structure. The spoof-SPPs-based communication system will be able to use the bidirectional end-fire antenna.

SSPP transmission line on a groundless metal layer can be used to design a high-gain end-fire antenna. An elliptical director and the asymmetric radiation element of SSPP TL are combined to achieve high gain [4]. With this configuration, the operating bandwidth of 15.75–17.5 GHz can be achieved with a high gain of 10.5–12.1 dBi and a high efficiency of 95%. Figure 6.1a, b shows the prototype of the antenna and its simulated and measured S-parameters. Figure 6.2 shows the measured and simulated radiation pattern for different frequency ranges. It is useful in wireless communications due to its single-layer, high gain, and low-profile advantages.

An end-fire radiating traveling-wave antenna with a low profile and great efficiency that uses a spoof surface plasmon polaritons (SSPPs) transmission line is studied here. It has a broad operating frequency band because of the traveling-wave nature. The single-layer design and subwavelength field confinement of SSPPs

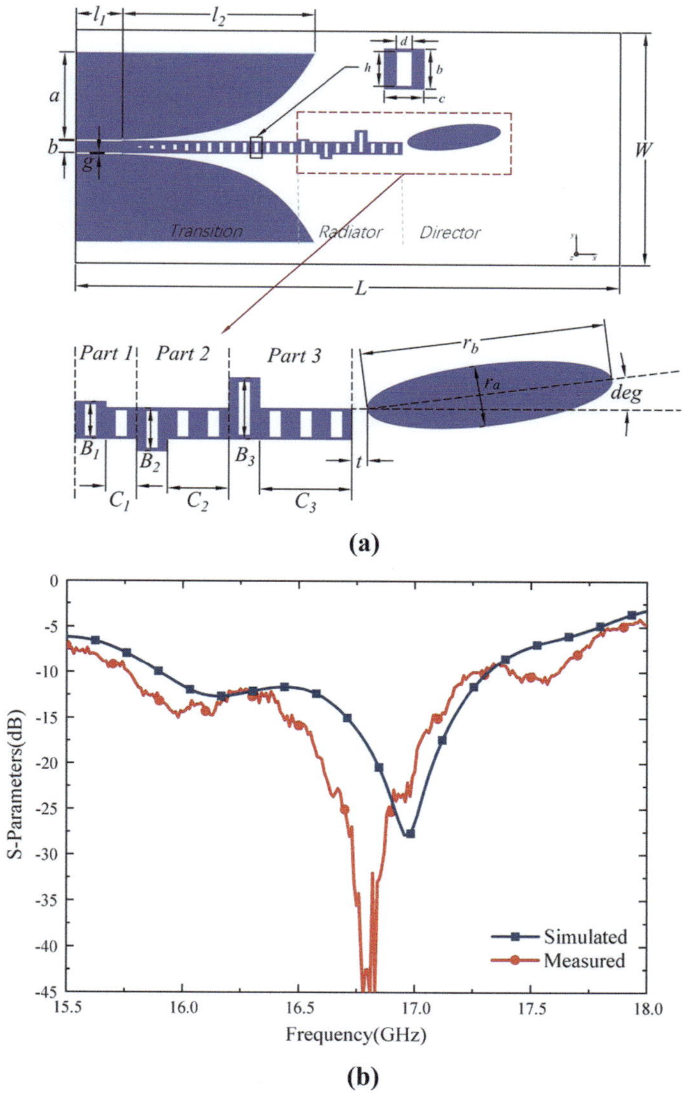

Fig. 6.1 **a** Prototype of the endfire antenna. **b** Simulated and measured S-parameter of the antenna [4]

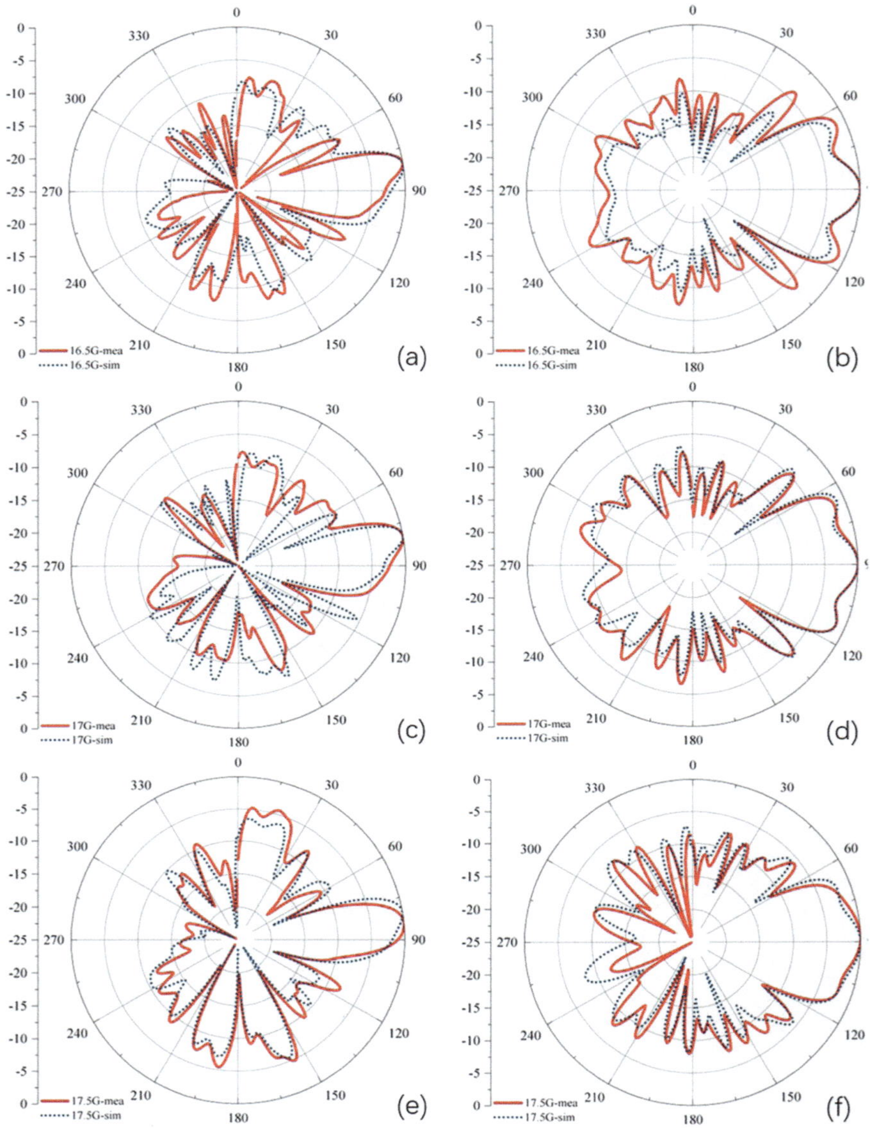

Fig. 6.2 Measured and simulated S-parameters of the end-fire antenna **a**, **c**, **e** xy-plane patterns at 16.5, 17 and 17.5 GHz, respectively. **b**, **d**, **f** xz-plane at 16.5, 17 and 17.5 GHz, respectively [4]

contribute to this antenna's low profile. Near-end-fire radiation beams at 7.5–8.5 GHz are produced by this antenna. 9.2 dBi is the maximum gain, and 96% is the total efficiency [5]. An end-fire filtering antenna can be designed using SSPP since filtering antennas are crucial for suppressing higher-order modes. Multiple radiating elements and a single double-sided parallel strip line (DSPSL) make up the end-fire leaky wave antenna's structure. Stubs with via holes are added to the antenna to achieve filtering performance. With a center frequency of 7.7 GHz and a bandwidth of 36.36%, the leaky wave end-fire filtering antenna based on SSPPs can achieve a peak gain of 8.65 dBi. More than 13 dB is rejected out of band [6].

Direction finding, tracking, and anti-interference applications are among the many uses for the end-fire antenna with sum and difference beams. When transmitting electromagnetic (EM) waves in SSPP modes, the asymmetric arrays of grooves with varying groove depths, periods, and/or duty ratios act as the transmission structure. The transformation of phase difference for the EM waves is realized because of the different phase constants for the EM fields on the two sides of the asymmetric structures. The end-fire radiations with sum and difference beams are formed by superposition in the end-fire direction for the in-phase and out-of-phase EM fields. The end-fire antenna's operating frequencies are 5.50, 6.60, and 7.09 GHz for the sum beam and 6.10 and 6.90 GHz for the difference beam [7].

The SSPP waveguide, which is created by etching a series of periodic corrugated grooves on the metal strip, has a high transmission performance that excites the antenna's main radiator. An extended U-shaped metal plane that serves as the antenna's reflector is placed in the bottom layer of the antenna in order to increase the end-fire antenna's front-to-back ratio. To increase directivity, three printed metal strips are intended to serve as the end-fire antenna's directors. At 7 GHz, the antenna's designed frequency, its gain is approximately 9.1 dBi, and its fractional bandwidth reach up to 31% [8]. Substrate-integrated waveguide (SIW) feeding structures and a triangle-shaped SSPP structure, which function as an array of tiny current sources, can be added to end-fire antennas to increase their performance. While the SIW's rectangular slots can increase bandwidth, the flared slot and tapered SSPP help us to improve gain and sidelobe levels. In the broad frequency range of 16–22.7 GHz, a maximum realized gain of 15.2 dB and high radiation efficiency of roughly 90% is attained [9].

The end-fire antenna can reach a wide bandwidth in the range of 4.1–8.1GHz owing to its composition of SSPP waveguides for the transmission line, a metal plate on the ground serving as the antenna's reflector, a metal strip director, and two half-rings to realize the radiation. Figure 6.3 shows photograph of fabricated end-fire antenna (a) Top view, (b) Bottom view, (c) Simulated and measured S-parameter of the end-fire antenna, and (d) Realized gain and efficiency of the end-fire antenna. This antenna operates in a broad frequency range, from 4.1 to 8.1 GHz, with a gain of 6.5 dBi, a bandwidth of 65%, and an efficiency of 97% [10]. By combining a fishbone-shaped structure for horizontal polarization (HP) and a mushroom-shaped structure for vertical polarization (VP), a dual-polarized end-fire antenna operating in the Ka-band is created. This planar antenna supports two orthogonally polarized surface wave modes. It produces a wide bandwidth and high-gain radiation. For HP and VP, the peak realized gain is 16.9 and 16.0 dBi, respectively. For HP and VP, the measured 3 dB gain bandwidth is 36.3 and 24.4% [11].

A dual-sided, tapered, symmetrical spoof surface plasmon polariton (SSPP) transmission line (TL) with tilted unit cells is used to design a planar antenna with dual-linearly polarized end-fire radiation. There are two first high-order modes of operation

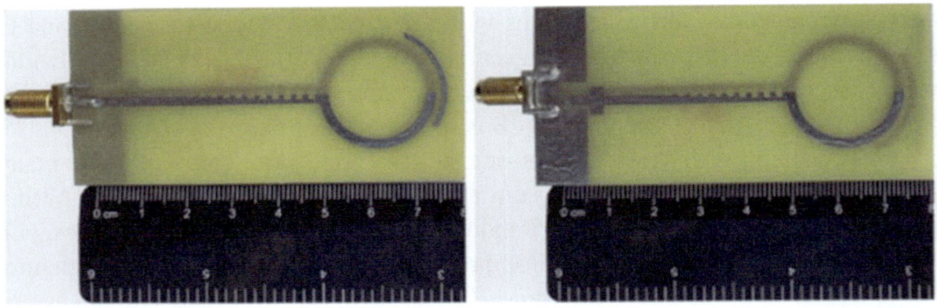

(a) (b)

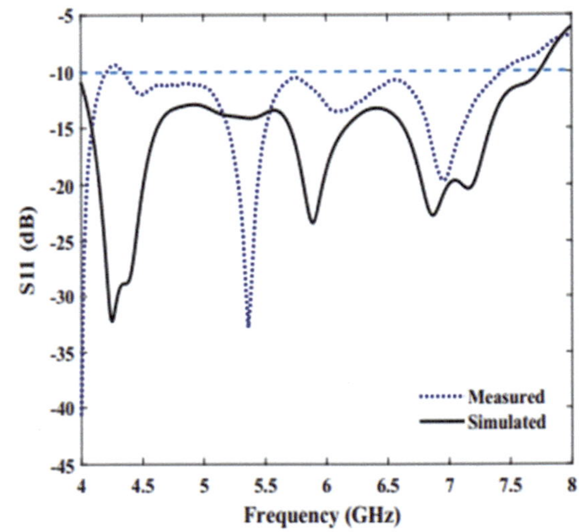

(c)

Fig. 6.3 Photograph of the fabricated end-fire antenna. **a** Top. **b** Bottom. **c** Simulated and measured S-parameter of the end-fire antenna. **d** Realized gain and efficiency of the end-fire antenna [10]

for the SSPP TL: the first high-order even mode and the first high-order odd mode. To feed the SSPP TL, a two-port feeding structure is intended. The energy from the signal fed from port 1 passes through the substrate-integrated waveguide and ultimately activates the SSPP TL's even mode, producing vertically polarized radiation. On the other hand, when the signal enters through port 2, the energy passes through the stripline-to-slotline transition before stimulating the odd mode on the SSPP TL and producing radiation that is horizontally polarized [12]. Figure 6.4a shows photograph of the fabricated dual polarized end-fire antenna and Fig. 6.4b shows simulated S-parameter of the end-fire antenna. Figure 6.5 shows the 3D radiation pattern of the antenna (a) Vertical polarization at 18 GHz (b) Vertical polarization at 26 GHz (c) Horizontal polarization at 18 GHz (d) Horizontal polarization at 26 GHz. A 54.6%

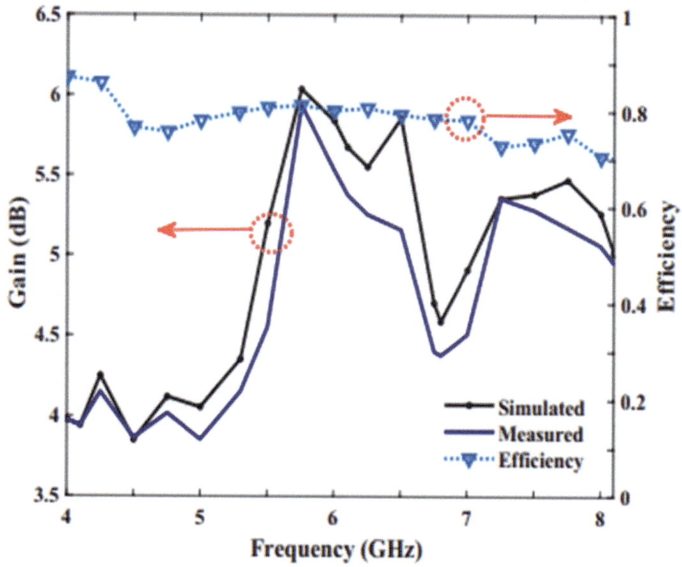

(d)

Fig. 6.3 (continued)

overlapping bandwidth is obtained. The horizontal and vertical polarisations achieve a maximum gain of 14.7 and 12.2 dBi, respectively.

The advantage of a flexible end-fire antenna based on spoof surface plasmon polaritons (SSPPs) at the Millimeter-wave band is that it can guarantee that the antenna's field distribution remains nearly constant when it is bent, resulting in relatively stable impedance characteristics. Next, a symmetrical SSPP transmission line fed by a microstrip-slotline structure is proposed to realize the end-fire radiation. Last but not least, an end-fire antenna has an average gain of 14.9 dBi over the frequency range of 30–50 GHz. Additionally, this antenna can still maintain good radiation performance when bent, and the bending angle changes linearly with its maximum radiation angle [13]. The flexible antenna exhibits good stable radiation performance when its bending angle varies from 0° to 30°, its maximum radiation angle increases from 0° to 24°, and its gain fluctuates within 2 dB. The antenna's benefits include a large bandwidth, compact size, and flexibility. It also has a promising future in wearable technology and flexible circuits.

Vertically polarized radiation is produced using a double-sided SSPP transmission line (TL). The band gap between the first and second high-order modes is removed by substituting a step-impedance-resonator (SIR) shaped stub for the uniform stub found in conventional SSPP cells. The maximum radiation point is oriented toward the end-fire direction due to the tilting of the double-sided SIR-shaped SSPP cells. To improve antenna gain and produce end-fire radiation, dielectric covers with air vias are added to the top and bottom surfaces of the SSPP TL. The radiator is fed by

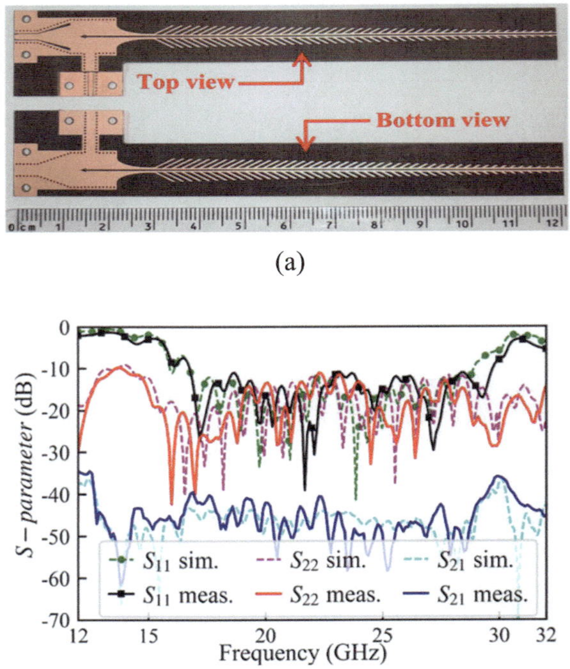

(a)

(b)

Fig. 6.4 **a** Photograph of the fabricated dual polarized end-fire antenna. **b** Simulated S-parameter of the end-fire antenna [12]

a double-sided parallel strip line transition via a grounded coplanar waveguide. With a peak gain of 17.3 dBi, the antenna attains a fractional bandwidth of 91.6% between 12.9 and 34.6 GHz [14]. To achieve a wideband high-efficiency end-fire radiation, a wideband end-fire antenna based on a modified antipodal Vivaldi radiator is designed using a double-layer SSPP transmission line with periodically etched arc slots and optimized transitions. Throughout the broad frequency range of 3 to 13 GHz, the antenna exhibits good end-fire radiation performances with $S_{11} < -10$ dB and an average efficiency of 95.9%. At 11 GHz, the average realized gain is 7.2 dBi, and the maximum realized gain is 9.13 dBi [15].

In [16] substrate-integrated waveguides (SIWs), a textured structure is incorporated into the effective surface plasmon polaritons (ESPPs) to excite the mode of effective spoof surface plasmon polaritons (ESSPPs), allowing for independent control of cut-off frequencies. A wideband filter is developed by taking advantage of the ESSPPs mode's transmission properties. Unlike ESPPs filters, each layer of an ESSPP has a single cut-off frequency determined by its effective permittivity, giving passband and out-of-band control independently. Moreover, it is advantageous to achieve compactness because the skirt selectivity of the filter can be improved without lengthening the filter. A filtenna is created by integrating the filter with a Vivaldi antenna.

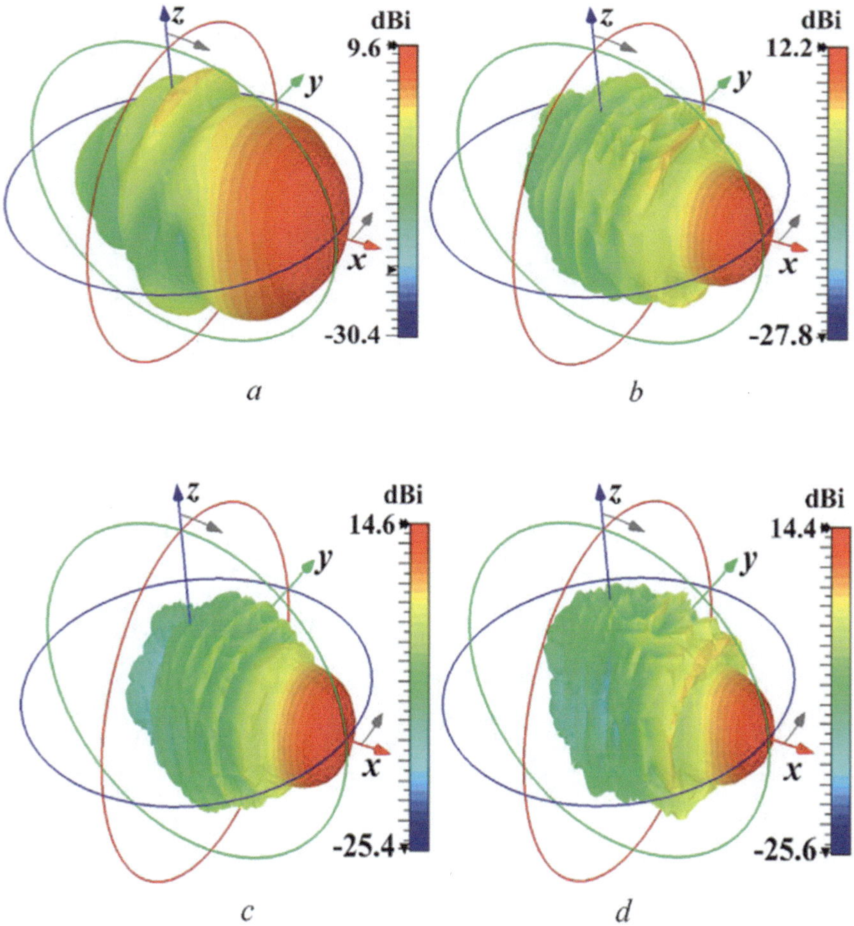

Fig. 6.5 3D radiation pattern of the antenna. **a** Vertical polarization at 18 GHz. **b** Vertical polarization at 26 GHz. **c** Horizontal polarization at 18 GHz. **d** Horizontal polarization at 26 GHz [12]

Based on spoof surface plasmon polaritons (SSPPs), this wideband antenna has a low profile, a simple structure, and transmission lines made of periodically modulated corrugated metal strips, a CPW feed, and an antenna reflector made of a ground metal plate. Figure 6.6 shows the fabricated antenna structure and the simulated and measured S-parameter of the antenna. The antenna operates in a broad operating band from 7 to 45 GHz, with a gain of 10.19 dB, bandwidth of 146%, and efficiency of 90%. Figure 6.7 shows the simulated and measured S-parameter of the end-fire antenna. High efficiency, good directivity, high gain, wide bandwidth, and ease of manufacturing are just a few of the design's many promising qualities [17].

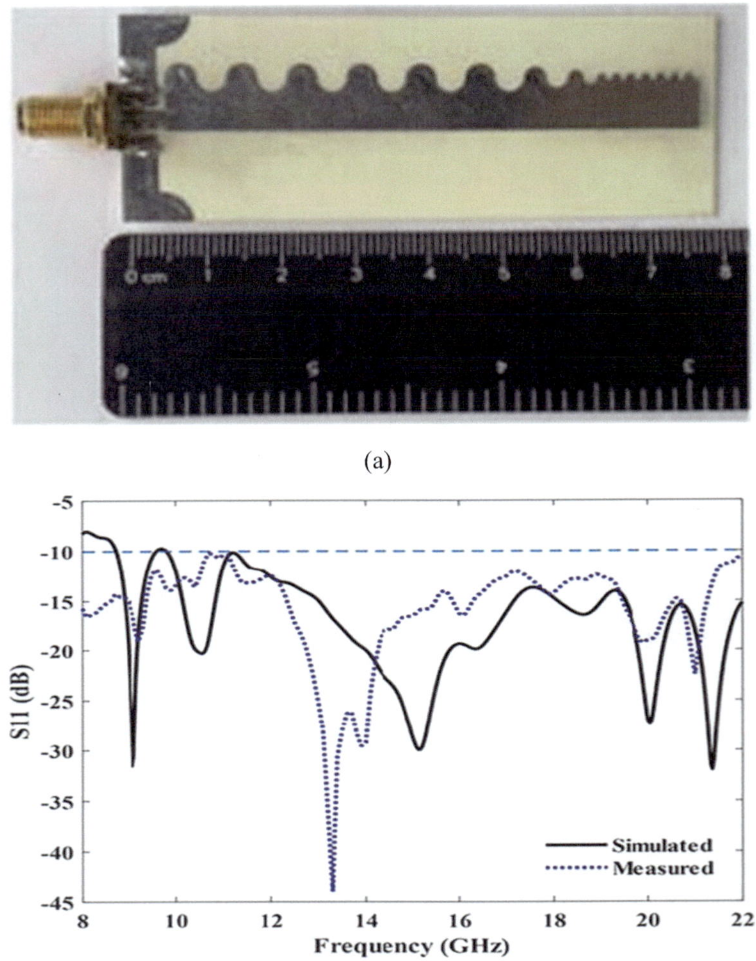

(b)

Fig. 6.6 **a** Fabricated end-fire antenna. **b** Simulated and measured S-parameter of end-fire antenna [17]

An easy way to design a new kind of end-fire antenna with a wideband, simultaneous bidirectional, and unidirectional radiation pattern. Two radiator structures connect directly to the two ports of a proposed splitter structure, which is used to excite different frequency bands that spoof surface plasmon polariton (SPP) modes. This allows for bidirectional and unidirectional radiation patterns in different frequency bands by dispersion mismatching, which decouples the spoof SPP modes from free space waves. This designed SPP spoof antenna has a bidirectional end-fire radiation pattern from 8.4 to 21.6 GHz and a unidirectional end-fire radiation pattern

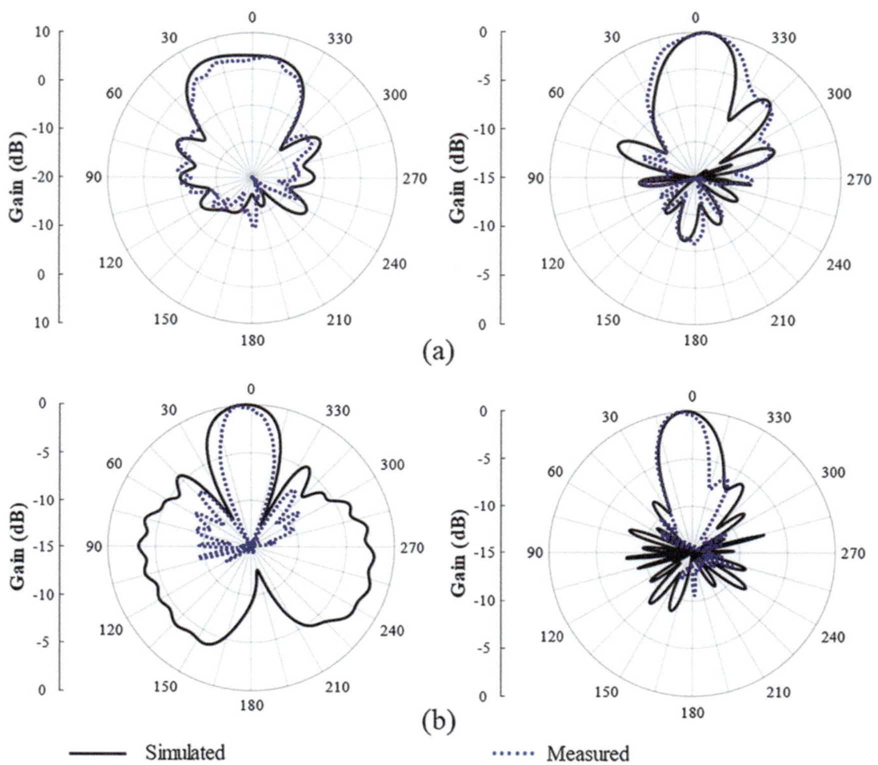

Fig. 6.7 Simulation and measurement of the radiation fields of plane E and plane H. **a** 6 GHz. **b** 6.8 GHz [17]

from 21.6 to 45.0 GHz. With a relative bandwidth of 137%, this designed spoof SPP antenna has an average gain of over 10.6 dB throughout its working bandwidth [18].

Two symmetrical SSPP transmission lines (TLs) are applied to an end-fire antenna to create a high aperture efficiency end-fire antenna based on odd-mode spoof surface plasmon polaritons (SSPPs). An 180° phase shifter balun supports the SSPPs TLs, enabling each pair of SSPPs units to function as a dipole and achieving an end-fire antenna. Then, to obtain the maximum gain at the end-fire direction, the Hansen–Woodyard condition is employed. Additionally, using these two SSPP TLs, we create a tapered structure that is capable of exciting radiation into free space. Consequently, an end-fire antenna with a zero tilted angle gets an average gain of 11.51 dBi and operates in the 9–11 GHz frequency range [19]. A high gain, planar end-fire filtering antenna based on spoof surface plasmon polaritons (SSPPs) with a controllable pass-band. Its radiating portion is connected to the end of the SSPP feeding line by three pairs of sector-shaped patches and I-shaped directors. By adjusting the size of all the branches with metal vias, and round-shaped, and π-shaped resonators, the proposed antenna's filtering performance can be readily modified. A level of out-of-band suppression exceeding 12 dB is attained along with an average gain of 8.6 dBi [20].

A two-conductor structure is used to design an end-fire frequency scanning tapered slot antenna (FSTSA), with a gradient slots array etched on only one of the antenna's two inner edges. The antenna can simultaneously excite two modes with distinct phase constants along both inner edges thanks to the asymmetric inner edges. The main lobe of the FSTSA can scan through the end-fire direction because of the dispersion characteristics of spoof surface plasmon polaritons (SSPPs), which cause the phase difference between the EM waves along the two conductors to chanas a function of frequencies. The antenna operates at a scanning rate of 26.4°/GHz, scanning from −15° to 14° (end-fire direction at 0°) between 6.60 and 7.70 GHz [21].

The feeding structure of this wideband, fishbone-shaped end-fire antenna can excite an odd-mode signal on the surface plasmon polaritons (SSPPs) using odd-mode spoof SPPs, which are made up of a differential-mode exciter and a microstrip-to-slotline converter. End-fire radiation, or the same as a row of tiny current sources, can be realized by the fishbone-shaped SSPP radiator. To achieve a wider upper bandwidth, a tapering end is added to the SSPP radiator. A broad impedance bandwidth of 70.3% (2.4–5.0 GHz) is achieved when $|S_{11}|$ is less than −10 dB as shown in Fig. 6.8. In the desired frequency band, low cross-polarization below −20 dB and stable end-fire radiation patterns are also achieved. Radiating efficiency is 90% and the gain ranges from 4.4 to 8.6 dBi over the whole operating band [22]. An end-fire antenna that satisfies the requirement of end-fire radiation by using asymmetric spoof surface plasmon polaritons (SSPPs) and phase reversal technique. In the meantime, the phase reversal technique gives the antenna a high gain. It has the highest end-fire gain of all single-layer designs, with a peak gain of 16.2 dBi in the operational frequency range of 17–19 GHz as shown in Fig. 6.9 [23].

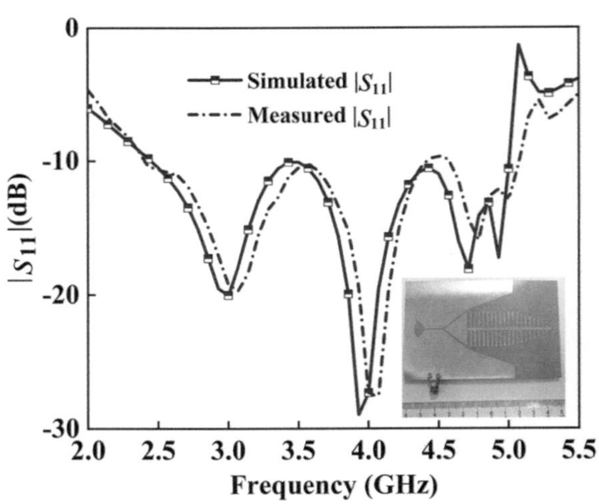

Fig. 6.8 S-parameter of end-fire antenna [22]. Reproduced with permission from the IEEE

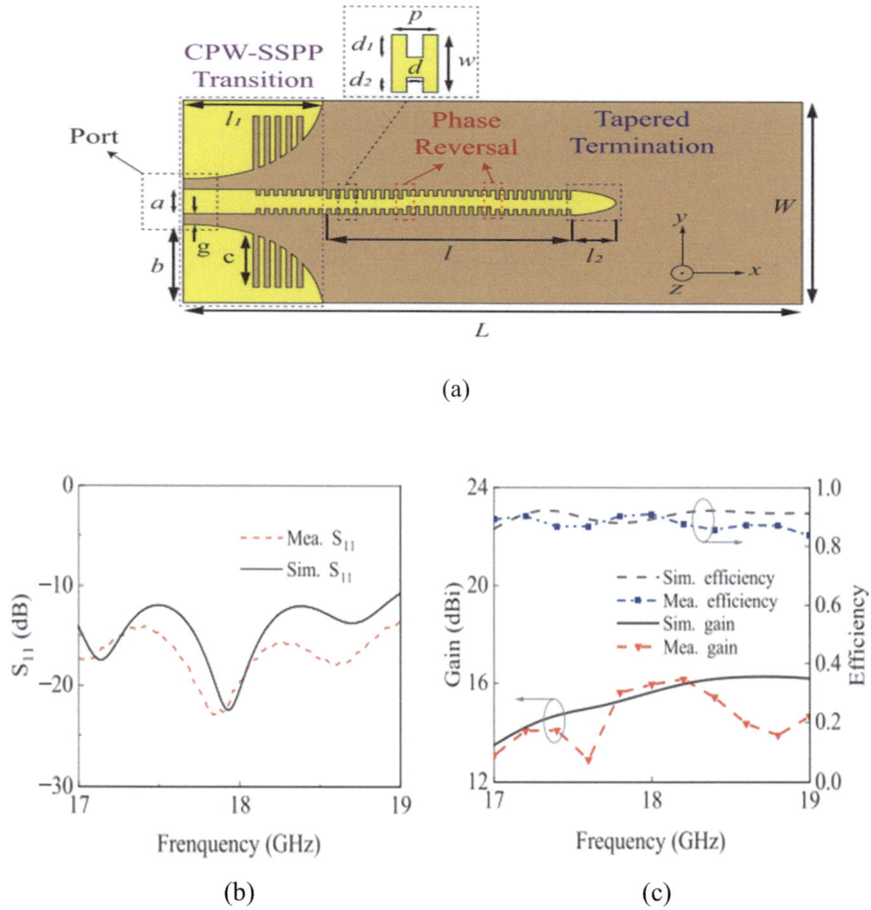

Fig. 6.9 **a** Schematic of end-fire antenna. **b** S-parameter of end-fire antenna. **c** Gain and efficiency plot of end-fire antenna [23]. Reproduced with permission from the IEEE

Dipole-shaped unit cells provide support for a wideband end-fire antenna based on odd-mode spoof surface plasmon polaritons (SSPPs). The typical antenna is divided into two half-length SSPP sections that form a two-element array since the introduction of phase reversal crossover. The extra phase provided by the phase reversal crossover improves the end-fire directivity of the suggested modified SSPPs antenna shown in Fig. 6.10. From 4 to 8 GHz, it achieves a relative bandwidth of 67%. 7.9 dBi is the maximum end-fire gain that can be achieved [24].

A planar end-fire antenna with high gain and wide bandwidth uses a dielectric lens and odd-mode spoof surface plasmon polaritons (SSPPs). An innovative asymmetric coplanar waveguide feeding configuration, comprising a row of gradient slots etched inside a single ground plane, is used to excite the SSPPs of the odd mode. To achieve stable end-fire radiation, the symmetric SSPP structure acts as a radiator. A planar

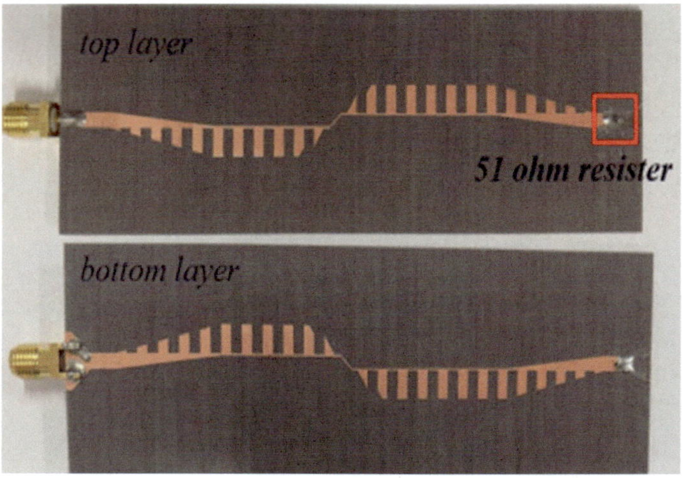

(a)

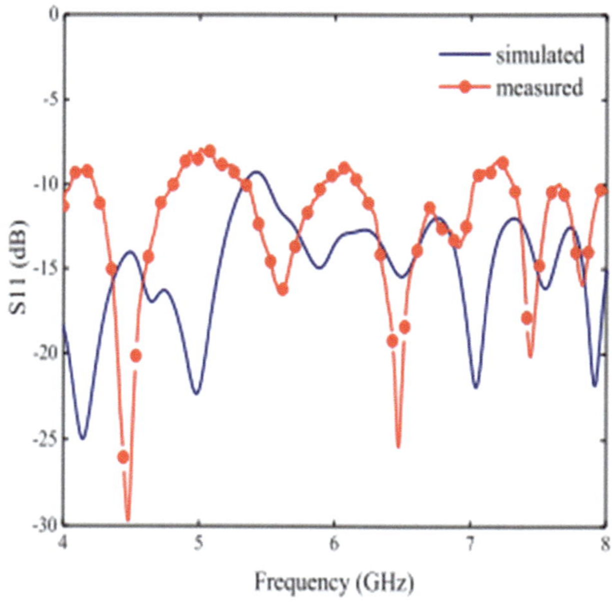

(b)

Fig. 6.10 **a** Fabricated prototype of end-fire antenna. **b** S-parameter of end-fire antenna [24]. Reproduced with permission from the IEEE elliptical dielectric lens is extended at the end of the substrate to improve the antenna's gain and front-to-back ratio (FBR). The end-fire antenna achieves a high peak end-fire gain of 14.8 dBi, an average efficiency of 80%, and the desired operating band of 7.9–13 GHz. Furthermore, over the entire bandwidth, the FBR is better than 20 dB, with an ideal peak FBR of up to 40 dB [25].

A surface plasmon polariton (SPP) bidirectional end-fire antenna operating at ultrawideband spoof wavelengths by mismatching wave vectors. Two types of slot lines are used to design the two radiator structures and the splitter structure that joins them to form the antenna. While the spoof SPP can communicate from the splitter structure to the two radiator structures, the spoof SPP's wave vectors in the splitter structure and radiator structures are out of alignment. To achieve a bidirectional end-fire radiation pattern, the spoof SPP from the splitter structure will not be in bounded mode when it propagates at the two radiator structures and will instead be decoupled by them. The working band of the splitter structure and the radiator structures' dispersion determine the decoupling bandwidth of the spoof SPP, which in turn determines the bandwidth of the proposed antenna. This antenna has an average efficiency of 93%, an average gain of more than 9.2 dB, and a relative bandwidth of 120% from 9.6 to 40.00 GHz [26].

Two independent surface waves with different nonlinear phase constants are supported by an asymmetric structure in a spoof surface plasmon polariton (SSPP) antenna with dual-band end-fire gain and a small frequency ratio (FR). Dual-band end-fire gains and end-fire radiation null are achieved at the terminal of the proposed structure because of the various phases of the superposed E-field. Further studies are conducted on the odd and even mode feedings, which offer extra phase differences. There are three designed prototypes with FRs of 1.06, 1.11, and 1.5. All of the results support the mechanism and show that by varying the unit cells and feeding modes, the suggested antenna's frequency response (FR) can be readily reached in the 1.06–1.5 range. Furthermore, a good directional interference inhibition is presented by the end-fire out-of-band gain suppression, which is over 17 dB [27]. A metal strip's double sides feature two symmetrically corrugated surfaces that make up an SSPP end-fire antenna. Two L-shaped slots and a power divider make up the feeding network, which simultaneously produces identically normal and oppositely tangential electric fields to excite the desired beam. Furthermore, flared feeding slot transitions excite the two magnetic current sources above the corrugated surfaces that are equivalent and tangential. Consequently, an artificial end-fire beam with a zero tilted angle is produced. With a 50% fractional bandwidth and a maximum gain of 16 dBi, the operating frequency range is 30–50 GHz [28].

The low gain and ultrawide bandwidth of the classic Vivaldi antenna are well known. Spoof surface plasmon polaritons (SSPP) are introduced as the parasitic element in an additional structure to improve the radiation characteristics. Without reducing bandwidth, the suggested patch offers an improved gain across the whole frequency range. A 5 dB improvement in the maximum gain is possible. Figure 6.11 shows the photograph of the fabricated antenna (a) top view (b) back view (c) shows the simulated and measured S-parameter of the antenna and (d) shows the realized gain of the antenna. With this design approach, the radiation characteristics of the

Vivaldi antenna can be improved in a novel way [29]. The design of an ultrawideband planar end-fire antenna is based on the first four high-order spoof surface plasmon polaritons (SSPPs) modes. The high-order modes are radiated via an SSPP transmission line with a progressively shallower groove depth. Additionally, the quad-band antenna's SSPP units are tilted to increase gain and sidelobe levels at high frequencies. The next step is to design an SSPP structure made of a periodic array of metal strips. The quad-band antenna's first two bandgaps are removed by adding two arrays of tilted I-shaped strip resonators. A fractional bandwidth of 140.8% is attained by the antenna between 6.9 and 39.7 GHz. 6.9 and 38 GHz yield minimum and maximum measured gains of 6.4 and 15.2 dBi, respectively [30].

An extensive beam steering angle is provided by a dual wideband antenna, each band featuring frequency scanning radiation and end-fire radiation. With a lower cut-off frequency and greater field confinement than a single-layer (S-L) spoof surface plasmon polaritons (SSPPs) structure, the antenna makes use of a double-layer (D-L) spoof SPP structure. While maintaining transmission in the slow wave region, the modulated D-L SSPP structure can excite the space harmonic distribution in the fast-wave region. It is possible to activate both the fast-wave (frequency scanning mode) and slow wave (end-fire mode) radiation in the two operating bands by combining the modulated D-L SSPPs waveguide and the Vivaldi radiator. In the two operating

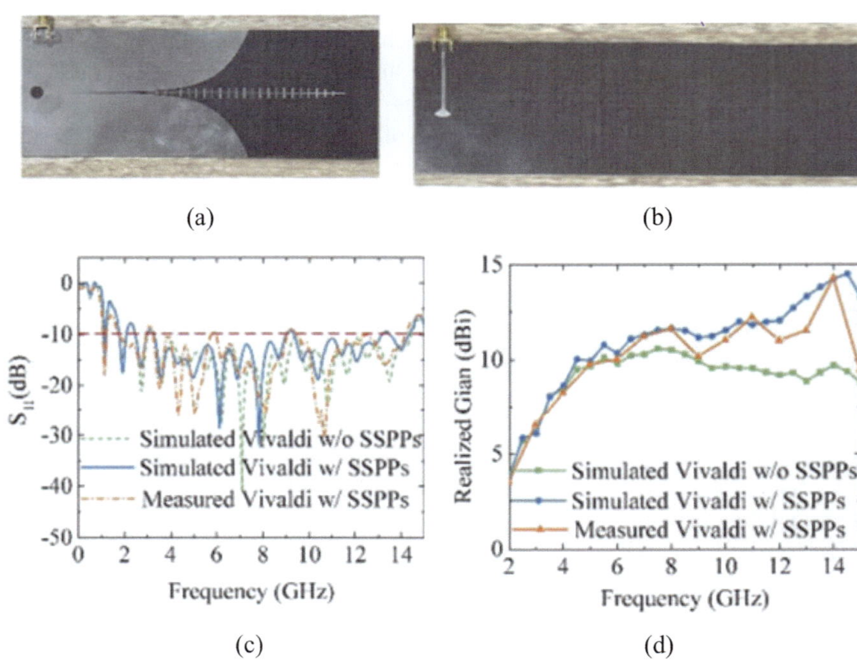

Fig. 6.11 Photograph of the fabricated antenna **a** top view **b** back view **c** simulated and measured S-parameter of the antenna **d** realized gain of the antenna [29]. Reproduced with permission from the IEEE

bands, it is possible to stimulate both the slow wave radiation (end-fire mode) and the fast-wave radiation (frequency scanning mode). The antenna operates in the end-fire mode with a beam direction of 90° and a realized gain that varies from 3.55 to 8.24 dBi in the slow wave range of 3–9 GHz. The antenna operates in the frequency scanning mode within the fast waveband from 13 to 34 GHz, achieving beam scanning from −90° to ±42.5° with a realized gain of 9.45–15.67 dBi [31].

6.3 Circularly Polarized End-Fire Antenna

Insensitive to polarization locations, immune to ionosphere-induced signal Faraday rotation, and highly anti-interference are just a few benefits of using circularly polarized (CP) antennas. They are thus widely employed in systems such as GPS, radar, and satellite communication. In this section, we will study a different type of circularly polarized end-fire antennas. The SSPP-based circularly polarized end-fire antenna is a great research area; in this topic limited research has been carried out and has a great scope. In [11] two orthogonally polarized surface wave modes are supported by this planar antenna, which is a dual-polarized end-fire antenna operating in the Ka-band. It is created by combining a mushroom-shaped vertical polarization (VP) structure and a fishbone-shaped horizontal polarization mechanism. Next, a seamless transition between the antenna and free space is accomplished by tapering these structures. It consequently attains a wide bandwidth and high gain radiation. Peak realized gain is reached, with HP and VP achieving 16.9 and 16.0 dBi, respectively. 32.4% for VP and 36.3% for HP is the 3-dB gain bandwidth.

Figure 6.12a shows the schematic of end-fire antenna with electric field plots for both horizontal and vertical polarization, whereas Fig. 6.12b shows the simulated and measured S-parameter of end-fire antenna for both horizontal and vertical polarization. Spoof surface plasmon polaritons coupled line is the basis for the design of an end-fire circularly polarized (CP) antenna. Spoof surface plasmon polariton (SSPP) structures act as a coupled line and a radiator when they are positioned on two orthogonal substrates. The circular polarization requires polarization orthogonality and phase quadrature, which this structure provides. Between 8.8 and 9.6 GHz, the return loss is more than 15 dB, and the axial ratio remains stable below 3 dB. At 9 GHz, the peak gain is 8.14 dBi. This structure can be fed by a single coaxial connector, significantly reducing the amount of space required for the feeding structure and its design complexity [32]. Two linear polarization (LP) antennas with varying strip heights are arranged orthogonally to create a circular polarized (CP) end-fire antenna. A feed monopole and ultrathin corrugated metallic strips make up a linear polarized (LP) end-fire antenna based on spoof surface plasmon polaritons (SSPP). The guided wave is coupled into the SSPP, propagates through the SSPP, and radiates out into free space via the metallic strips. The antenna's gain is proportional to the length of the strip structure, just like traveling wave antennas [33].

The feeding terminal, transition section, and asymmetric surface plasmon polaritons (SSPPs) structure make up a Millimeter-wave circularly polarized (CP) end-fire

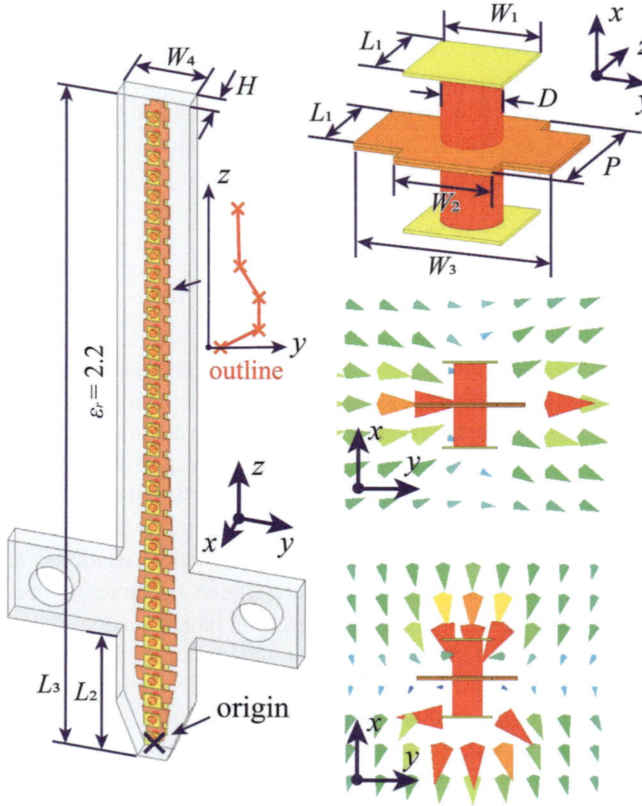

(a)

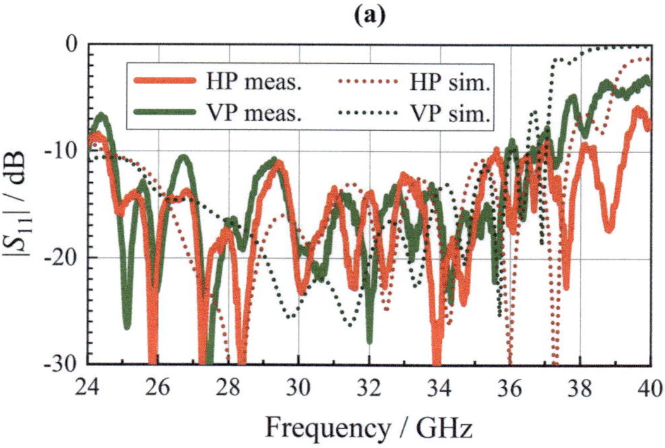

(b)

Fig. 6.12 **a** Schematic of end-fire antenna. **b** S-parameter of end-fire antenna [11]. Reproduced with permission from the IEEE antenna based on a single-sided spoof surface plasmon polaritons (SSPPs) structure. The formation of a flared slot resembling a planar horn antenna results in end-fire radiation. CP waves are produced by the microstrip line on the bottom layer and the SSPPs structure on the top layer. Within the impedance bandwidth of 26.4–28.2 GHz and the axial ratio bandwidth of 26.4–28.5 GHz, the antenna achieves good end-fire radiation performance [34]. Figure 6.13a shows the fabricated structure of the end-fire antenna and Fig. 6.13b shows the simulated and measured S-parameter of the end-fire antenna. Whereas (c–d) shows the normalized directivity of the E-plane and H-plane respectively.

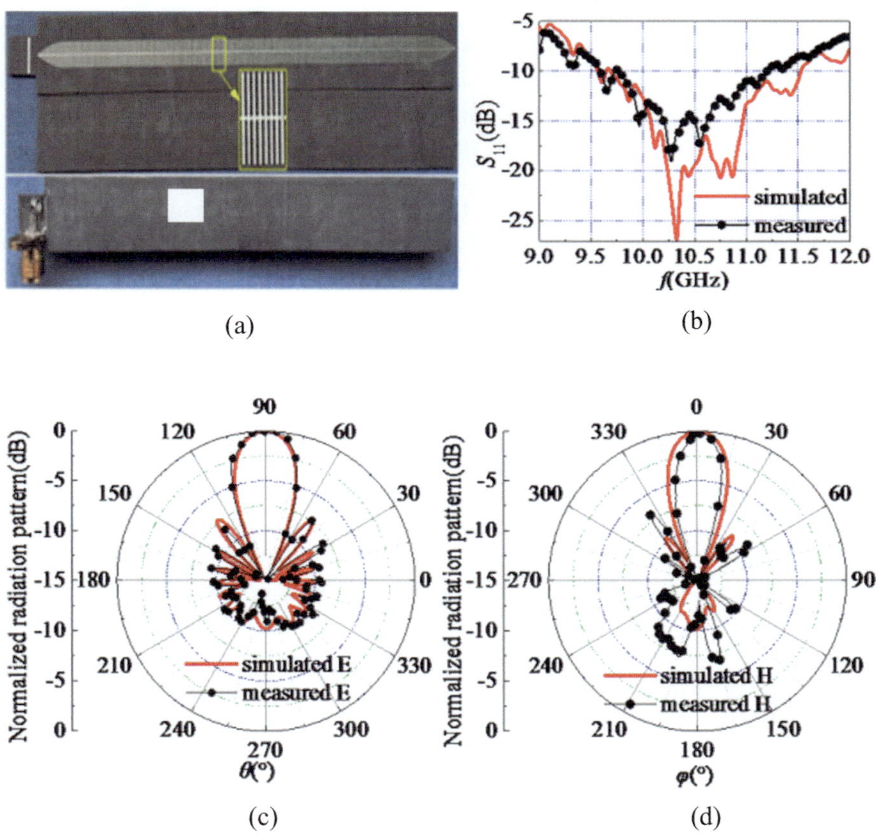

Fig. 6.13 **a** Fabricated structure of the end-fire antenna. **b** Simulated and measured S-parameters of the end-fire antenna. **c** Normalized directivity of E-plane. **d** Normalized directivity of H-plane [34]

6.4 Applications of End-Fire Antenna

Owing to their small size, low cost, and high gain, end-fire antennas are extensively utilized in various fields, such as long-distance communications. To enhance the capacity and prevent polarization mismatch in high-performance systems, a dual-polarized antenna is typically needed. Advantages like low profile, low complexity, depressed mutual coupling, and high signal integrity have been demonstrated when using the spoof surface plasmon polaritons-based end-fire antenna in MIMO configuration. It is also advantageous to miniaturize sophisticated circuits and systems because SSPPs-based structures can concentrate electromagnetic (EM) waves into subwavelength scales. Multiple-input-multiple-output (MIMO) applications can make use of the SSPP-based end-fire antenna [35]. The antenna consists of a coplanar waveguide (CPW)-fed SSPPs transmission line (TL), a transition to change the CPW quasi-TEM mode to the SSPPs mode, and a radiator section with asymmetric discontinuities to change the bounded SSPPs waves into radiating waves to radiate. Because of its excellent field confinement competency, the MIMO antenna takes advantage of the SSPPs TL's intrinsic high isolation feature. In the operating band of 9.52–11.23 GHz, wider beam widths are achieved for ample azimuth and elevation coverage with a stable pattern in the end-fire direction. Its maximum envelope correlation coefficient (ECC) is 1.5×10^{-3}, its total active reflection coefficient (TARC) value is less than 0.3, and its high isolation is measured at -35 dB in the operating range.

An aqueous glucose solution and body glucose measurement can be monitored using an SSPP end-fire antenna. Using a significantly smaller effective aperture, the sensor emits an end-fire beam into a glucose water solution. It has been demonstrated that sidelobes are ineffective in glucose sensing, so two triangular ground planes were added to the sensor's coplanar waveguide (CPW) port to suppress them. Measurements of glucose concentrations with a higher degree of sensitivity have been made possible by the slow wave characteristics of the suggested SSPP end-fire sensor. Measurements conducted on the body yield a high sensitivity of 150 MHz/mgmL^{-1} [36].

To create a compact broadband SSPP-based antenna that can precisely measure end-fire radiations through frequency scanning, asymmetric upper and lower corrugations and H-shaped periodic unit cells are used [37]. The proposed SSPP-based antenna's topology is depicted in Fig. 6.14. The CPW feeding section, the main body of the antenna, and the tapered impedance end comprise the three sections of the antenna. The lower and upper corrugated edges of the fabricated antenna form an array of identical H-shaped SSPP unit cells that serve as the antenna's guiding body, known as a Spoofed Surface Plasmon Polariton (SSPP) waveguide. Various

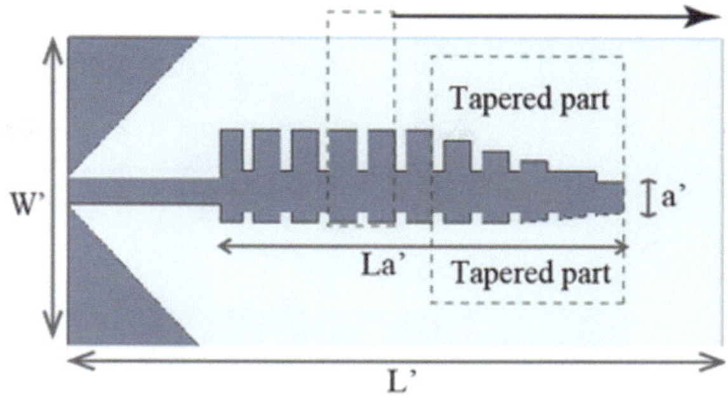

Fig. 6.14 Schematic of the end-fire antenna [37]

components of the manufactured design are shown with an explanation of their functions in Fig. 6.15a–d. The high field confinement required to achieve end-fire radiations is depicted in Fig. 6.15e. The final designed fabricated prototype is depicted in Fig. 6.15f. Here, we have carried out a preliminary experiment to demonstrate the usefulness of the design. The proposed SSPP design has three primary characteristics: (1) compact size; (2) efficient end-fire radiation; and (3) through end-fire scanning, which represents (4) multifunction capabilities of the antenna scanning application. The measured scattering parameters for the antenna are displayed in Fig. 6.16. The measured normalized radiation patterns of the antenna as a function of beam angle are displayed in Fig. 6.17a. The measured normalized radiation pattern of the antenna in the end-fire direction is displayed in Fig. 6.17b. As seen in Fig. 6.18, the target detection system consists of a spoof surface plasmon antenna, the object being tested, a measurement tool, and simulation software. The reflection parameters for the suggested detection system are displayed in Fig. 6.19. The reflection has been measured after the object under test (OUT) has been positioned at two distinct distances (D1 and D2). In this section, we explained two applications of end-fire antennas in communication and biomedical applications.

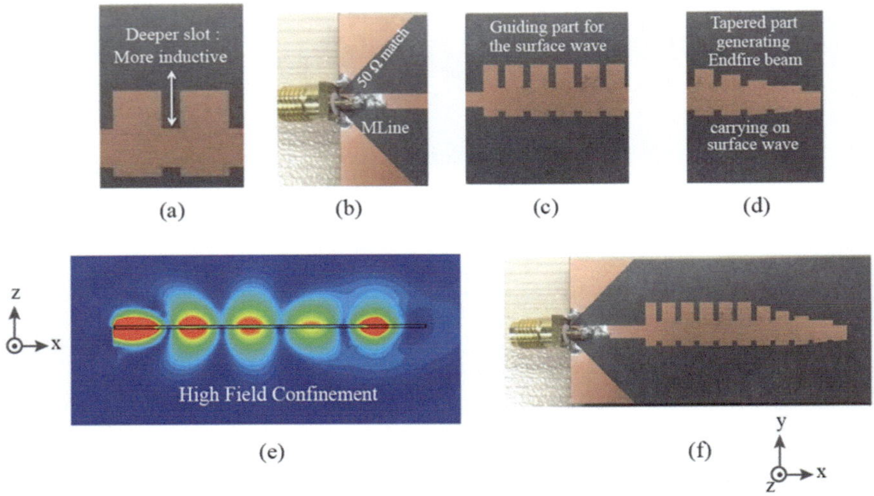

Fig. 6.15 Proposed SSPP antenna X **a** Unit element. **b** Transition part. **c** Guiding part of the SSPP. **d** Tapered part of the SSPP. **e** Fields along the axis cut plane. **f** Final prototype [37]

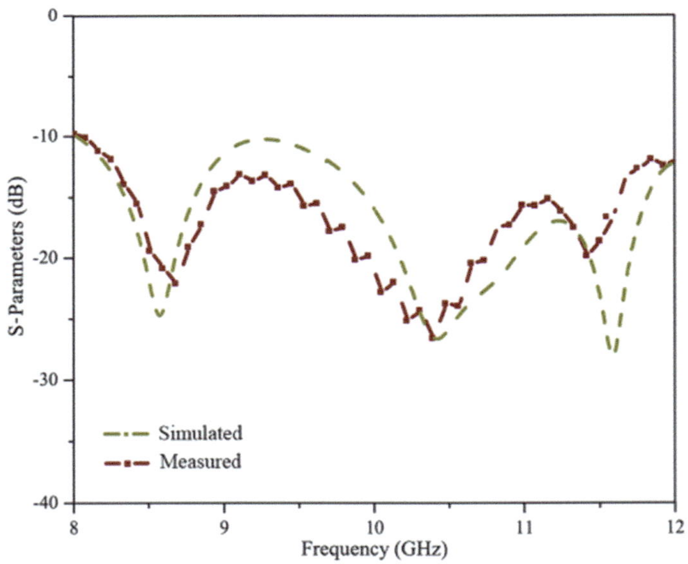

Fig. 6.16 Measured scattering parameters of SSPP antenna [37]

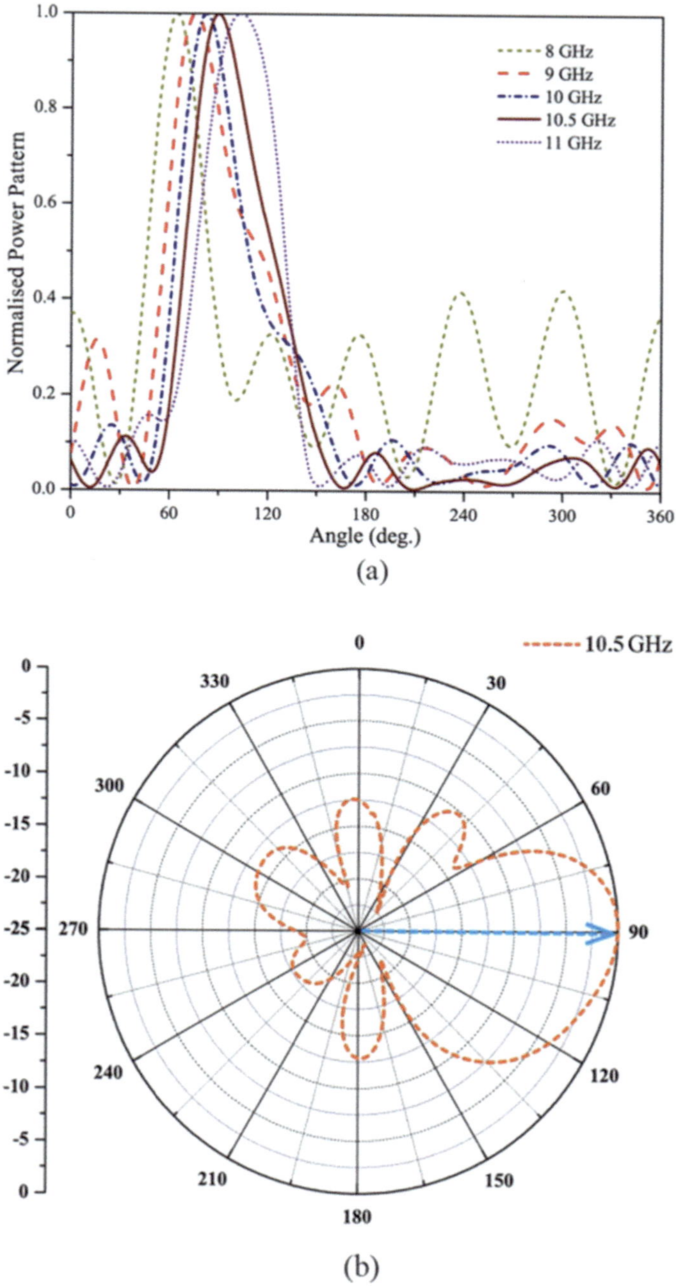

Fig. 6.17 **a** Measured scanning patterns of the SSPP antenna. **b** Measured end-fire radiation pattern of SSPP antenna [37]

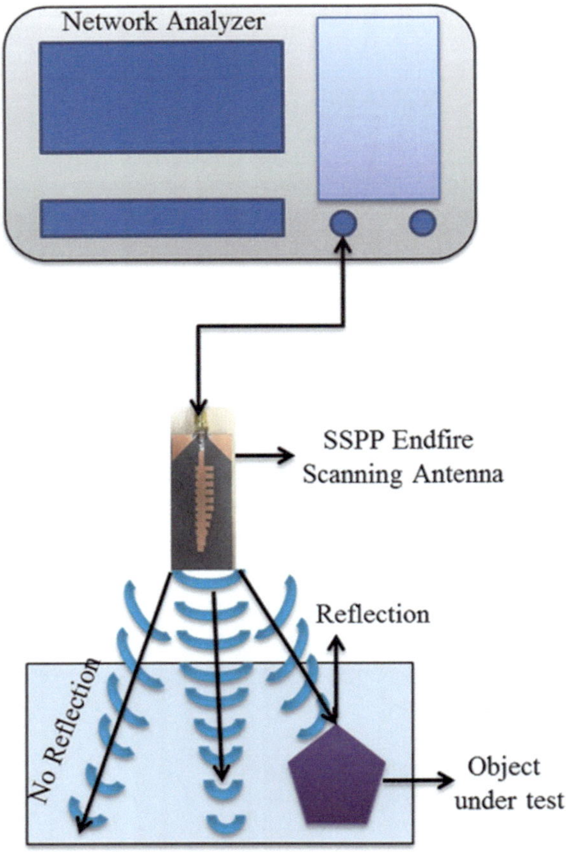

Fig. 6.18 Measurement setup for the target location using an SSPP antenna [37]

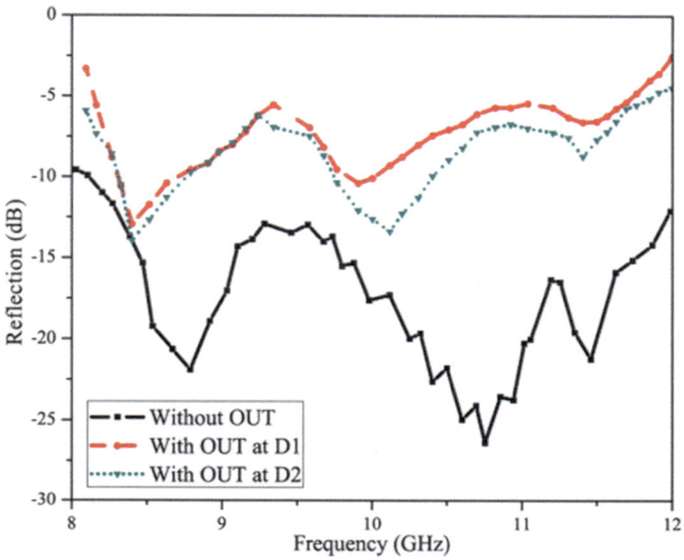

Fig. 6.19 Reflection measurements at different distances [37]

6.5 Conclusion

In this chapter, we studied different techniques to design end-fire antennas and their applications. The circular polarized end-fire antenna is studied in detail and its advantages in the antenna domain are highlighted. Further, the applications of end-fire antennas in the field of MIMO and the biomedical domain are discussed in detail. This chapter helps to understand the design procedure of end-fire antenna with different methodologies and helps to understand how to design a high gain, wideband, and low-profile end-fire antennas.

References

1. Yin, J.Y., et al.: Endfire radiations of spoof surface plasmon polaritons. IEEE Antennas Wirel. Propag. Lett. **16**, 597–600 (2017). https://doi.org/10.1109/LAWP.2016.2592512
2. Zhang, X.-F., Fan, J., Chen, J.-X.: High gain and high-efficiency millimeter-wave antenna based on spoof surface plasmon polaritons. IEEE Trans. Antennas Propag. **67**(1), 687–691 (2019). https://doi.org/10.1109/TAP.2018.2879847
3. Tian, D., et al.: Low-profile high-efficiency bidirectional endfire antenna based on spoof surface plasmon polaritons. IEEE Antennas Wirel. Propag. Lett. **17**(5), 837–840 (2018). https://doi.org/10.1109/LAWP.2018.2818109
4. Liu, L., Chen, M., Yin, X.: Single-layer high gain endfire antenna based on spoof surface plasmon polaritons. IEEE Access **8**, 64139–64144 (2020). https://doi.org/10.1109/ACCESS.2020.2984153

5. Kandwal, A., Zhang, Q., Tang, X.-L., Liu, L.W., Zhang, G.: Low-profile spoof surface plasmon polaritons traveling-wave antenna for near-endfire radiation. IEEE Antennas Wirel. Propag. Lett. **17**(2), 184–187 (2018). https://doi.org/10.1109/LAWP.2017.2779455
6. Feng, W., Feng, Y., Wu, L.-S., Shi, Y., Zhou, X.Y., Che, W.: A novel leaky wave endfire filtering antenna based on spoof surface plasmon polaritons. IEEE Trans. Plasma Sci. **48**(9), 3061–3066 (2020). https://doi.org/10.1109/TPS.2020.3013608
7. Li, S., Zhang, Q., Xu, Z., Zhao, H., Yin, X.: Phase transforming based on asymmetric spoof surface plasmon polariton for endfire antenna with sum and difference beams. IEEE Trans. Antennas Propag. **68**(9), 6602–6613 (2020). https://doi.org/10.1109/TAP.2020.2993083
8. Yang, L., Xu, F., Jiang, T., Qiang, J., Liu, S., Zhan, J.: A wideband high-gain endfire antenna based on spoof surface plasmon polaritons. IEEE Antennas Wirel. Propag. Lett. **19**(12), 2522–2525 (2020). https://doi.org/10.1109/LAWP.2020.3038492
9. Homayoon, F., Heidari, A.-A., Pezhman, M.-M.: A low-profile high-performance endfire antenna based on spoof surface plasmon polaritons. IEEE Antennas Wirel. Propag. Lett. **22**(12), 2891–2895 (2023). https://doi.org/10.1109/LAWP.2023.3303638
10. Arghandeh, F., Abbasi-Arand, B., Hesari-Shermeh, M.: A wideband and compact Quasi-Yagi antenna based on spoof surface plasmon polaritons. Sci. Rep. **13**, 11054 (2023)
11. Pan, Y., Dong, Y.: Dual-polarized printed endfire antenna based on spoof surface plasmon polariton. IEEE Antennas Wirel. Propag. Lett. **22**(5), 1129–1133 (2023). https://doi.org/10.1109/LAWP.2023.3234571
12. Liang, Z., Qi, S.-S., Wu, W.: Dual-polarised end-fire antenna based on the high-order mode of spoof surface plasmon polaritons. IET Microw. Antennas Propag. **18**(1), 52–58 (2024)
13. Fu, Q., Liu, L., Luo, G.Q.: A flexible millimeter-wave endfire antenna based on spoof surface plasmon polaritons. IEEE Antennas Wirel. Propag. Lett. **23**(1), 29–33 (2024). https://doi.org/10.1109/LAWP.2023.3315521
14. Liang, Z., Qi, S.-S., Wu, W.: Wideband vertically polarized planar endfire antenna based on spoof surface plasmon polaritons. IEEE Antennas Wirel. Propag. Lett. **22**(9), 2240–2244 (2023). https://doi.org/10.1109/LAWP.2023.3281840
15. An, G., Zhu, C., Zhang, Y., Li, W., Jiang, Y., Ye, L.: A wideband endfire antenna incorporating modified antipodal vivaldi resonator and spoof surface plasmon polaritons. IEEE Antennas Wirel. Propag. Lett. **23**(6), 1874–1878 (2024). https://doi.org/10.1109/LAWP.2024.3372191
16. Zhang, A., Wang, J., Yi, Y., Liu, L.: Effective spoof surface plasmon polariton-enabled independent cutoff frequency control in wideband endfire vivaldi filtenna. IEEE Trans. Antennas Propag. **72**(4), 3073–3080 (2024)
17. Arghandeh, F., Abbasi-Arand, B., Hesari-Shermeh, M.: A compact wideband antenna with high gain based on spoof surface plasmon polaritons. Sci. Rep. **14**, 10107 (2024)
18. Dong, G., Li, X., Liu, Y., Zhang, A.: Spoof surface plasmon polariton endfire antenna possessing bidirectional and unidirectional radiation pattern. IEEE Antennas Wirel. Propag. Lett. **21**(1), 69–73 (2022). https://doi.org/10.1109/LAWP.2021.3118442
19. Fu, Q., Ni, H., Luo, G.Q., Zhu, L., Liu, L.: A high aperture efficiency endfire antenna based on spoof surface plasmon polaritons. IEEE Trans. Antennas Propag. **71**(1), 50–57 (2023). https://doi.org/10.1109/TAP.2022.3213417
20. Liu, L., Fu, Q., Hu, Y., Luo, G.Q., Zhu, L.: An endfire controllable-filtering antenna based on spoof surface plasmon polaritons. IEEE Antennas Wirel. Propag. Lett. **21**(3), 526–530 (2022). https://doi.org/10.1109/LAWP.2021.3137613
21. Zhao, H., Ge, F., Zhang, Q., Li, S., Yin, X.: Asymmetric endfire frequency scanning tapered slot antenna with spoof surface plasmon polaritons. IEEE Trans. Antennas Propag. **70**(7), 5913–5917 (2022). https://doi.org/10.1109/TAP.2022.3140505
22. Du, X., Li, H., Yin, Y.: Wideband fish-bone antenna utilizing odd-mode spoof surface plasmon polaritons for endfire radiation. IEEE Trans. Antennas Propag. **67**(7), 4848–4853 (2019). https://doi.org/10.1109/TAP.2019.2913707
23. Xiao, M., Rashid, A.K., Liu, B., Fan, R., Zhang, Q.: Design of high-gain single-layer endfire antenna using phase-reversed asymmetric spoof surface plasmon polaritons. IEEE Antennas Wirel. Propag. Lett. **22**(3), 641–644 (2023)

24. Jiang, Y., Liu, L., Hu, Y., Jiang, D.: Wideband small aperture endfire antenna based on spoof surface plasmon polaritons. IEEE Trans. Antennas Propag. **69**(8), 5026–5031 (2021). https://doi.org/10.1109/TAP.2021.3060141
25. Xu, J., Zheng, Z.: High-gain and wideband planar endfire antenna implemented via spoof surface plasmon polaritons and dielectric lens for X-band applications. IEEE Antennas Wirel. Propag. Lett. **22**(2), 382–386 (2023). https://doi.org/10.1109/LAWP.2022.3213553
26. Dong, G., Liu, Y., Li, X., Zhang, A.: Ultrawideband spoof surface plasmon polariton bidirectional endfire antenna based on wave vectors mismatching. IEEE Trans. Antennas Propag. **69**(11), 7895–7899 (2021). https://doi.org/10.1109/TAP.2021.3076560
27. Zhao, H., Li, J., Zhang, Q., Li, S., Yin, X.: Spoof surface plasmon polariton antenna with dual-band endfire gain and flexible small frequency ratio. IEEE Trans. Antennas Propag. **70**(11), 11079–11084 (2022). https://doi.org/10.1109/TAP.2022.3184499
28. Liu, L., Jiang, Y., Hu, Y., Jiang, D., Zhu, L.: Wideband millimeter-wave endfire antenna based on symmetrical spoof surface plasmon polaritons. IEEE Trans. Antennas Propag. **69**(11), 7386–7393 (2021). https://doi.org/10.1109/TAP.2021.3076199
29. Liao, Z., Che, Y.Z., Luo, G.Q., Zhang, Z.H., Qian, Y.H., Cai, B.G.: Enhanced radiation characteristics for Vivaldi antenna using spoof surface plasmon polaritons. IEEE Trans. Plasma Sci. **49**(9), 2730–2736 (2021). https://doi.org/10.1109/TPS.2021.3101814
30. Liang, Z., Qi, S.-S., Wu, W.: Ultra-wideband planar end-fire antenna based on high-order modes of spoof surface plasmon polaritons. IEEE Trans. Antennas Propag. **71**(2), 1469–1477 (2023). https://doi.org/10.1109/TAP.2022.3227750
31. Zu, H.-R., Wu, B., Zhao, Y.-T., Su, T., Fan, Y.-F.: Dual-band antenna with large beam steering angle incorporating endfire and frequency scanning modes using double-layer SSPPs structure. IEEE Trans. Antennas Propag. **70**(1), 46–55 (2022). https://doi.org/10.1109/TAP.2021.3096556
32. Zhao, F., Zhang, K., Zhang, Q., Li, S., Zhao, H., Yin, X.: Endfire circularly polarized antenna based on spoof surface plasmon polaritons coupled-line. In: 2022 IEEE 10th Asia-Pacific Conference on Antennas and Propagation (APCAP), Xiamen, China, pp. 1–2 (2022). https://doi.org/10.1109/APCAP56600.2022.10069519
33. Huang, J., Shao, Y., Zhang, C., Zhang, J.: A circularly polarized endfire antenna based on single-sided spoof surface plasmonpolaritons. In: 2022 International Symposium on Antennas and Propagation (ISAP), Sydney, Australia, pp. 459–460 (2022). https://doi.org/10.1109/ISAP53582.2022.9998634
34. Han, Y., et al.: Endfire antennas based on spoof surface plasmon polaritons. In: 2017 International Applied Computational Electromagnetics Society Symposium (ACES), Suzhou, China, pp. 1–2 (2017)
35. Jatav, R., Mali, R., Rathore, P.S., Meshram, M.K.: A low-profile high-isolation endfire MIMO antenna based on spoof surface plasmon polaritons for X-band applications. AEU Int. J. Electron. Commun. **177**, 155233 (2024). ISSN 1434-8411. https://doi.org/10.1016/j.aeue.2024.155233
36. Kandwal, A., et al.: Surface plasmonic feature microwave sensor with highly confined fields for aqueous-glucose and blood-glucose measurements. IEEE Trans. Instrum. Meas. **70**, 1–9 (2021)
37. Kandwal, A., Li, J., Igbe, T., et al.: Broadband frequency scanning spoof surface plasmon polariton design with highly confined endfire radiations. Sci. Rep. **10**, 113 (2020)

Chapter 7
Spoof Surface Plasmon Polariton-Based Power Dividers and Phase Shifters

7.1 Introduction

In the previous chapters, we have already discussed different types of transitions, transmission lines, filters, and antennas based on spoof surface plasmon polariton. It can be observed that the advantages of this technology over conventional technology make it an important candidate for component design in the field of microwave, millimeter wave, and terahertz frequency domains. In this chapter, we will discuss the design of some of the basic components that are very important in the field of microwave, millimeter, or terahertz domain. We will discuss spoof surface plasmon polariton-based components such as power dividers and phase shifters in detail.

Power divider plays an important role in all the circuits as they are the basic component to split power. Power dividers can be of many types like simple T-type power dividers which are used to divide power but do not have good isolation.To overcome this issue a Wilkinson power divider is designed which helps to improve isolation between the outport ports with the use of a simple resistor.

The phase shifter is another important element in microwave circuits such as a phased array antenna where the phase shifter helps to switch the beam in different directions. Phase shifters can be designed for different bit configurations. Spoof surface plasmon polariton-based phase shifters make it advantageous as by varying the corrugation, different phase variations can be achieved. Compared to conventional technology the SSPP-based phase shifters are compact.

7.2 Power Divider

Spoof surface plasmon polariton (SSPP) beam splitter is the simplest configuration of the power divider where a compact Y-shaped beam splitter integrated with a broadband rejection filter [1]. Single-band, multiple bands as well as broadband

S. K. Koul and S. Sharma, *Microwave Integrated Circuits Based on Spoof Surface Plasmon Polariton*, Signals and Communication Technology,
https://doi.org/10.1007/978-981-96-9041-1_7

rejections are achieved by introducing different symmetric T-shaped decorations. All these configurations can be achieved by using different configurations of T-shaped and rectangular grooves. It is possible to integrate multiple microwave functions in the same component, e.g., filtering antennas, filtering switches, multifunctional radiofrequency (RF) passive and active components capable of performing several RF analog signal-processing functions in the same circuit volume, and so on.

Power divider based on SSPP technology helps us to achieve ultra-wideband suppression of harmonics because of their low pass characteristics. Different types of SSPP transmission lines such as double-sided with bow-type grooves can be used to design a power divider. Figure 7.1a shows the fabricated prototype of the power divider [2]. Figure 7.1b shows the simulated and measured S-parameter of the power divider with a 10 dB return loss bandwidth of 146% from 0.9 to 5.73 GHz and the upper stop band can be extended up to 40 GHz with a suppression level of up to 32 dB. Ultra-wideband isolation can be achieved by using two lumped resistors between the two branches of the SSPP transmission lines of the power divider.

A circular hole instead of rectangular corrugation can be used to design a transmission line with a much more confined field. This type of transmission line can be used to design a power divider [3]. The S_{21} and S_{31} of the power divider are equal and about 3dB with the cut-off frequency of 10.8 GHz which means the input power is divided into two outports equally. As the isolation resistors are not used in this case the isolation between the ports is less than −10 dB.

Instead of conventional corrugated structures, Greek-cross fractal units (GCFUs) can be used to design an SSPP transmission line. The SSPP transmission line with

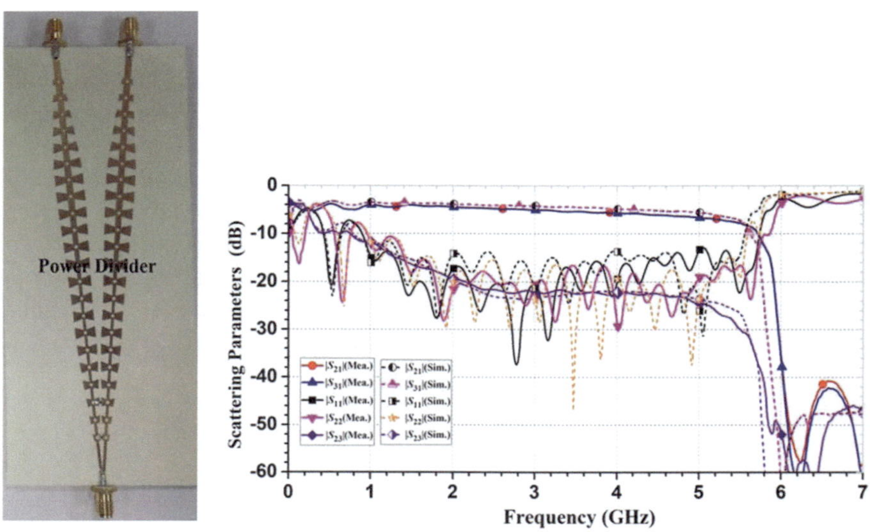

(a) (b)

Fig. 7.1 **a** Fabricated prototype of a power divider. **b** Simulated and measured S-parameter of the power divider [2] these fractal structures can be used to design an SSPP transmission line that has low pass characteristics like that of a conventional SSPP transmission line. A wideband Y-splitter includes an SSPP waveguide with double-sided GCFUs and two SSPP waveguide branches with single-sided GCFUs [4]. To improve isolations and reduce the reflections between output ports of the Y-splitter, a 100 Ω resistor is loaded in the middle of two symmetrical branches to construct a Wilkinson splitter to equally and efficiently split the energy of the SSPPs into two parts [4]. The simulated and measured results show that the SSPP splitter possesses excellent performance in a passband of 1.5–4.0 GHz with low insertion loss (S_{21}, $S_{31} > -4$ dB), high reflection loss (S_{11}, S_{22}, $S_{33} < -10$ dB), and high isolation ($S_{32} < -10$ dB). Wilkinson splitters have good transmission performance with low insertion loss, low bend loss, and excellent conformal capacity.

A Wilkinson power divider is one that has good isolation; an oval ring-shaped SSPP unit cell can be used to design the divider. A single resistor is placed in the middle of two symmetrical half oval rings to realize the isolation between the two output ports over the spectrum of 4.5–7.5 GHz. Figure 7.2a shows the schematic of the Wilkinson power divider whereas Fig. 7.2b shows the photograph of the fabricated Wilkinson power divider [5]. Figure 7.3a shows the simulated and measured $|S_{11}|$ and $|S_{21}|$ without a resistor, (b) shows simulated and measured $|S_{11}|$ and $|S_{21}|$ with a 200 Ω resistor, (c) shows simulate $|S_{22}|$ with and without a 200 Ω resistor, (d) shows measured $|S_{22}|$ with and without a 200 Ω resistor and (e) shows the simulated and measured $|S_{32}|$ with and without a 200 Ω resistor. The CPW to SSPP transition is used to design an efficient power flow between the input and SSPP transmission line. The stepped-impedances are co-designed with the three input/output ports of the power divider to achieve the impedance matching between the SSPPs waveguides and the new half oval ring-shaped SSPPs waveguide.

The combination of spoof surface plasmon polariton (SSPPs) and substrate integrated waveguide (SIW) results in a bandpass structure because of the properties of SSPP as a low pass filter and SIW as a high pass filter. The hybrid-SIW-SSPP structure has a good chance of increasing the device's bandwidth control range. All it takes to achieve bandwidth is to load interdigital structures (ISs) or optimize its dimensions. Additionally, on the bottom layer, defective ground structure (DGS) is added to enhance the out-of-band suppression even more. Furthermore, by welding additional resistors across the SSPP slot, good isolation can be achieved [6]. Figure 7.4a shows the photograph of the fabricated structure. The distribution of electric-field components along the z-axis is shown in Fig. 7.4b–d at 6, 10, and 13 GHz, respectively. A prototype measuring 45 mm by 18 mm by 0.5 mm is constructed and loaded with a dumbbell-shaped DGS and two resistors. S_{21} and S_{31} are located within the entire X-band at about −3.8 dB. The suggested method provides a workable way to decrease the size of devices based on SSPP and increase the bandwidth of devices based on SIW.

One of the most crucial power divider design factors is to achieve a different power divider ratio. Because SSPP transmission lines are less sensitive to the thickness, dielectric constant, and loss tangent of the selected substrate below the cut-off frequency than microstrip lines, they are a promising option. Various power divider

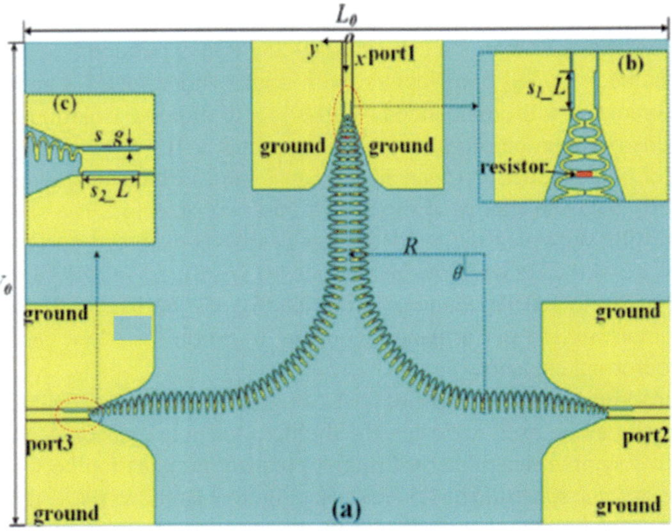

(a)

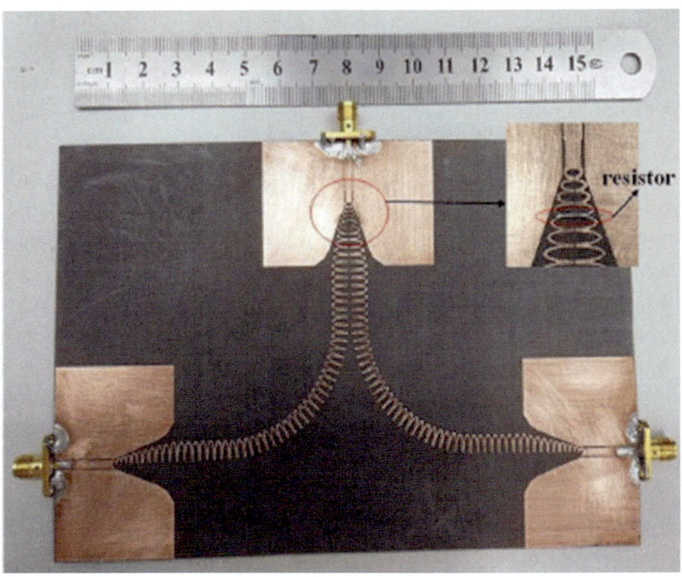

(b)

Fig. 7.2 **a** Schematic of power divider. **b** Photograph of the fabricated power divider [5]

Fig. 7.3 **a** Simulated and measured |S11| and |S21| without a resistor. **b** Simulated and measured |S11| and |S21| with a 200 Ω resistor. (c) Simulate |S22| with and without a 200 Ω resistor. **d** Measured |S22| with and without a 200 Ω resistor. **e** Simulated and measured |S32| with and without a 200 Ω resistor [5]

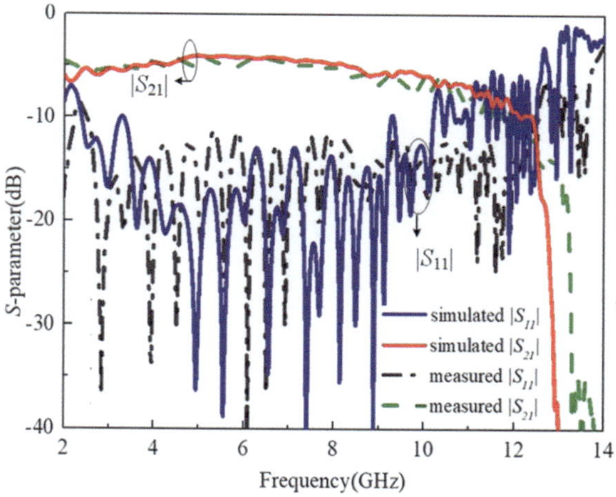

(a)

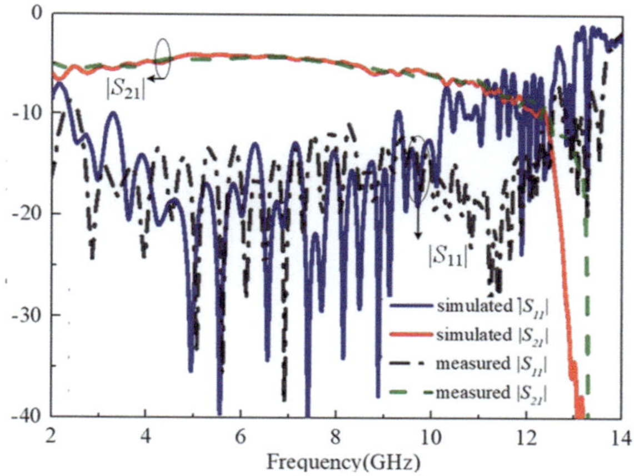

(b)

ratios of 0, 2, and 5 dB have many benefits, such as wideband operation, single-conductor coplanar quasi-symmetry structures, and the ability to design convenient implementations of various power-dividing ratios. Figure 7.5a shows the photograph of the fabricated SSPPs-based power dividers. Simulated and measured scattering parameters and the photograph of the power dividers with different power-dividing ratios are shown in Fig. 7.5b–d 0 dB equal power divider, 2 dB unequal power divider, and 5 dB unequal power divider, respectively, [7]. A versatile substitute for controllable output power division, the SSPPs-based unequal power divider will be crucial to the advancement and use of SSPPs-based circuits and systems at microwave and terahertz frequencies.

A 3-way ultra-wideband power divider based on spoof surface plasmon polariton can be designed using hexagonal ring unit cells [8]. The creation of electromagnetic

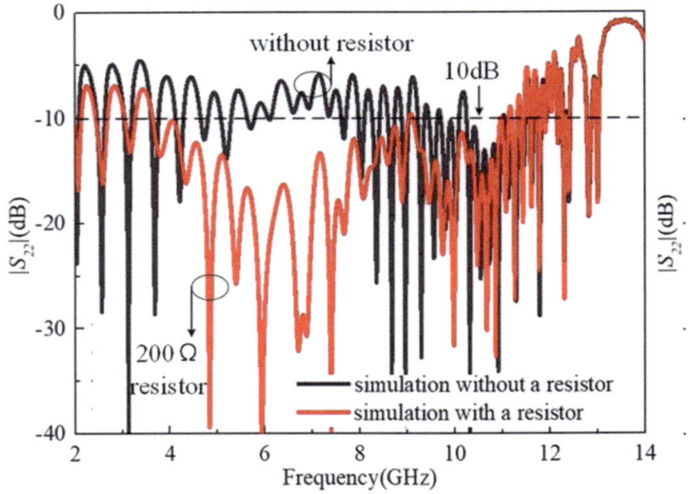

(c)

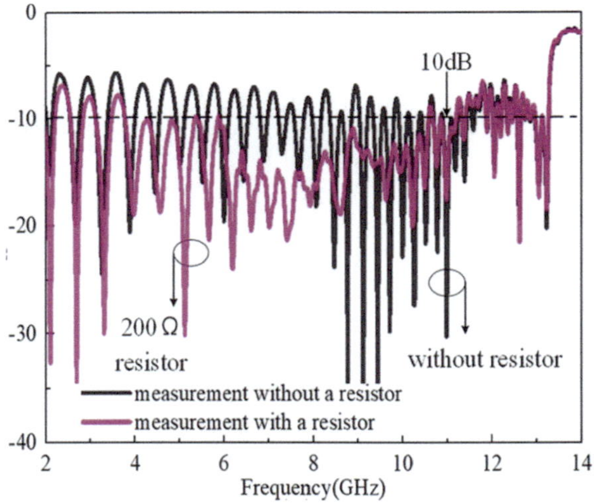

(d)

Fig. 7.3 (continued)

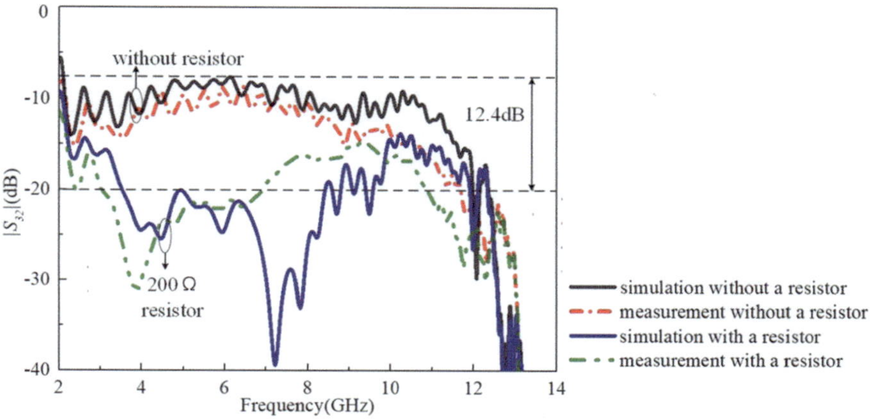

(e)

Fig. 7.3 (continued)

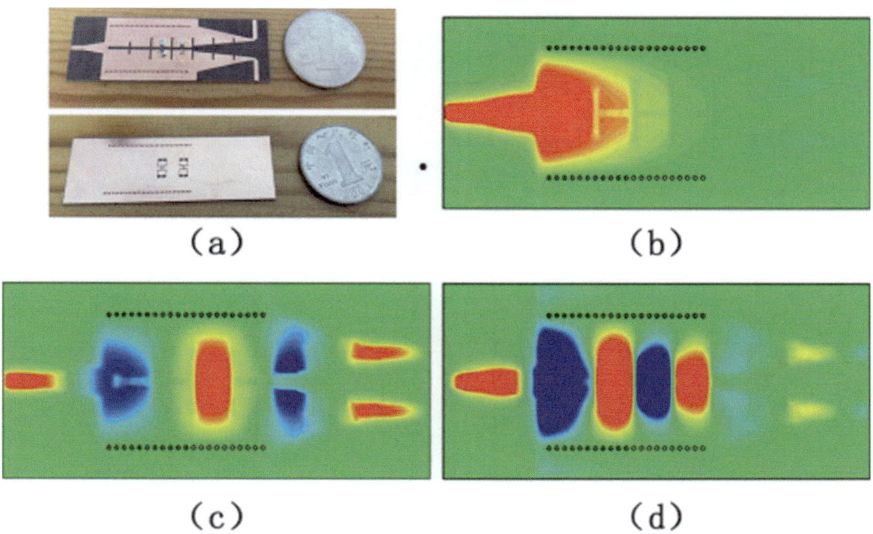

Fig. 7.4 **a** Photograph of the fabricated structure. Distribution of electric-field component along z-axis at **b** 6 GHz, **c** 10 GHz, and **d** 13 GHz [6]. Reproduced with permission from the IEEE

waves with high confinement in a collision between periodically decorated metal and dielectric is a fundamental property of SSPPs. An equal and in-phase 3-way power divider with good isolation between three output ports is designed in a wide frequency range of 2.5–8 GHz with an FBW of 105%. The ultra-wide-bandwidth, low loss, flexibility, and stable performance of this power divider prove its high capability compared with the state-of-the-art references. The n-way power divider could be utilized in integrated plasmonic RF and microwave circuits and systems,

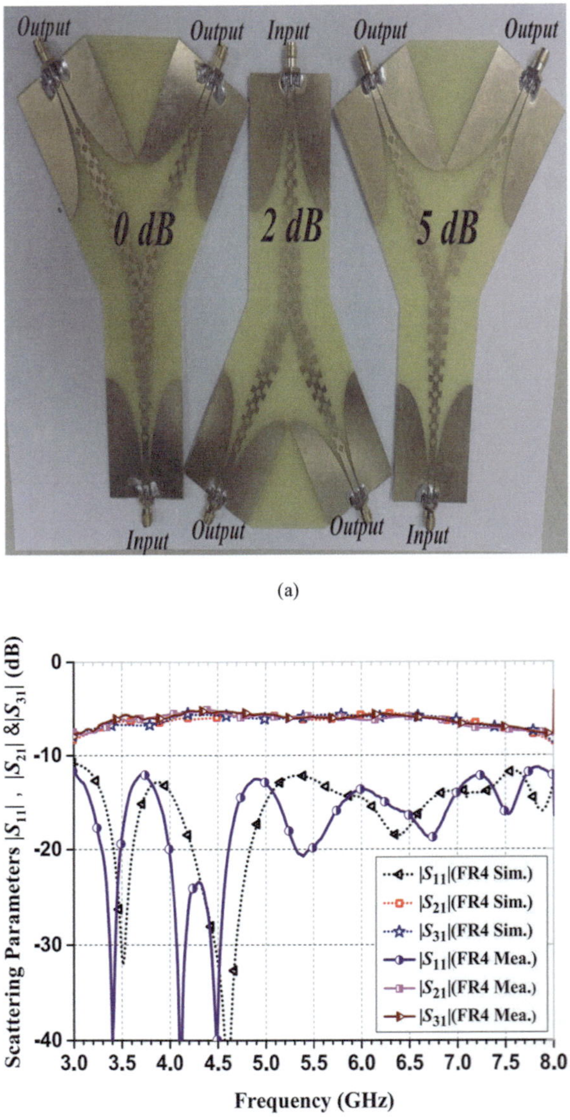

Fig. 7.5 **a** The photograph of the fabricated SSPPs-based power dividers. Simulated and measured scattering parameters and the photograph of the power dividers with different power-dividing ratios. **a** 0 dB equal power divider, **b** 2 dB unequal power divider, and **c** 5 dB unequal power divider [7]

such as balanced mixers, modulators, and the feed network of beamforming antenna arrays.

Reconfigurable or tunable circuits play an important role in the advancement of the power divider for future reconfigurable circuits. A tunable spoof surface plasmon transmission line can be achieved when the transmission line is periodically loaded with varactors and short-circuited stubs. Compared with the common SSP-TL with open-circuit stubs, the SSP-TL with varactor and short-circuited exhibits better

Fig. 7.5 (continued)

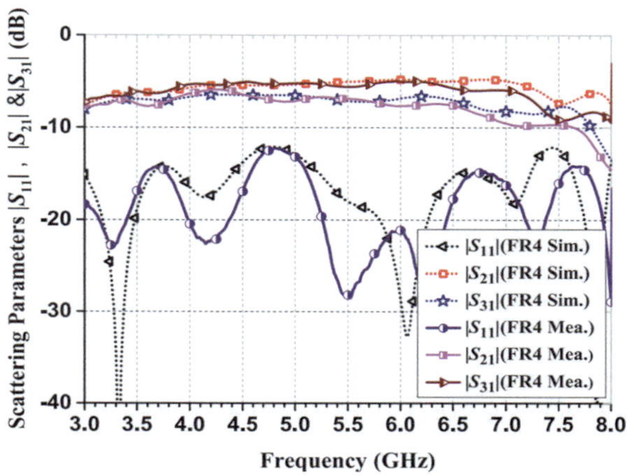

(c)

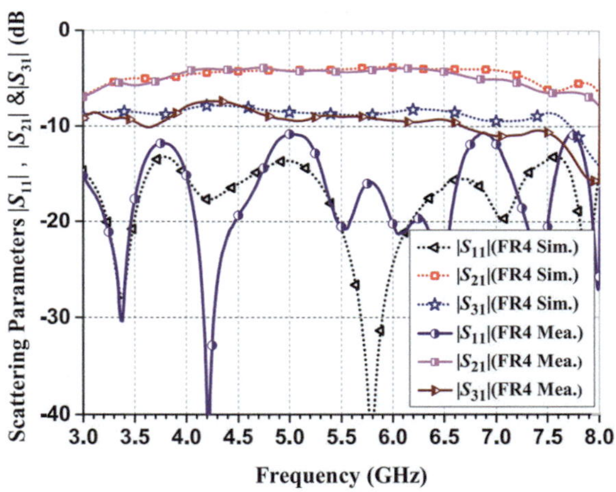

(d)

compactness and higher tunability. The SSP-TL can also be biased in a conventional microstrip bias circuit and seamlessly integrated into microstrip circuits [9]. As applications, a frequency-reconfigurable and compact Wilkinson power divider is designed and fabricated using the proposed SSP-TL as the λ/4 impedance transformer around 1 GHz. It is shown that 35% (or 20%) frequency tunability is achieved reducing the footprint to 46% of the conventional microstrip one.

A hybrid of half-mode substrate integrated waveguide (HMSIW) and spoof surface plasmon polariton (SSPP) is used in this C-band filtering power divider (FPD). With features from both the SSPP and HMSIW structures, the hybrid circuit is created by etching subwavelength corrugated grooves in the upper wall of the HMSIW [10]. It can produce the bandpass response and cut the substrate integrated waveguide structure's transverse and longitudinal dimensions by almost half. By adjusting the dimensions of the HMSIW and SSPP structures separately, the lower and upper edges of the passband can be tuned. With a minimum insertion loss of 1.05 dB and a maximum return loss of 10.5 dB, the bandwidth of 66% spans from 4 to 8 GHz.

Spoof surface plasmon polariton can be used to design different types of wideband power dividers using single and double-layer periodic spoof surface plasmon polaritons. A double-sided parallel-strip line is used to realize the wideband and low-loss T-shaped power dividers with large bandwidth, wideband isolation, and low loss. Figure 7.6a shows the photograph of the fabricated prototype of the wideband power divider. Figure 7.6b–d shows the measured and simulated results of the wideband power divider; (b) Transmission and insertion loss of the power divider, (c) Power division isolation and (d) Phase difference [11]. Conventional coplanar waveguide is used as the output ports for the single-layer SSPPs. SSPPs power dividers.

Based on the coupled mode theory, this compact broadband spoof surface plasmon polaritons (SPPs) power divider has good isolation and outstanding transmission efficiency. To divide the energy of spoof SPPs evenly between the two output ports into two equal parts with good isolation, an equal power divider is built [12]. Four coupling units, which are positioned close to the input port to divide energy between the two output ports, make up this coupling method. The coupling units and spoof SPPs are both used in the energy transfer process. Three ends of the spoof SPP waveguide are intended for signal feeding or extraction via coplanar waveguides with a 50 Ω impedance. The fields of SPPs along a single waveguide in a particular frequency range can easily be coupled into two adjacent branch waveguides with the same propagation constants, according to the coupled mode theory. This method has been used in a 3-port lossy network for impedance matching.

A novel modified split ring resonator (MSRR) integrated with a groove gap waveguide (GGW) and spoof surface plasmon polariton (SSPP) is used to design a millimeter-wave dual wideband full-metal filtering power divider (FPD) with low insertion loss. The first step in designing a single-wideband T-junction FPD is to embed SSPP into the GGW groove. It is convenient to adjust the lower and upper cut-off frequencies by varying the GGW width and SSPP height, respectively, [13]. Additionally, to enhance impedance matching, a tuning pillar is used. Moreover, an additional stopband is added to the single-wideband FPD by means of two coupled MSRRs, the location and size of which allow for nearly independent tuning of the stopband's center frequency and bandwidth. The frequency ranges of the lower and upper bands are 20.92–24.92 and 27.48–32.12 GHz with average insertion losses of 0.45 and 0.70 dB, respectively. Within both bands, the magnitude and phase differences between the two output ports are lower than 0.3 dB and 6°, respectively.

It has been possible to reconfigure the dispersion properties of spoof surface plasmon polaritons (SSPPs) by loading tunable components like varactors and p-i-n diodes [14]. Nonetheless, transmission loss of SSPPs will result from the parasitic resistance of tunable elements, a phenomenon that has not yet been thoroughly studied. Herein, a thorough response to the following three important questions is provided: (1) how to assess the loss of reconfigurable SSPPs; (2) how to lessen the loss of reconfigurable SSPPs; and (3) how to make use of the loss of reconfigurable

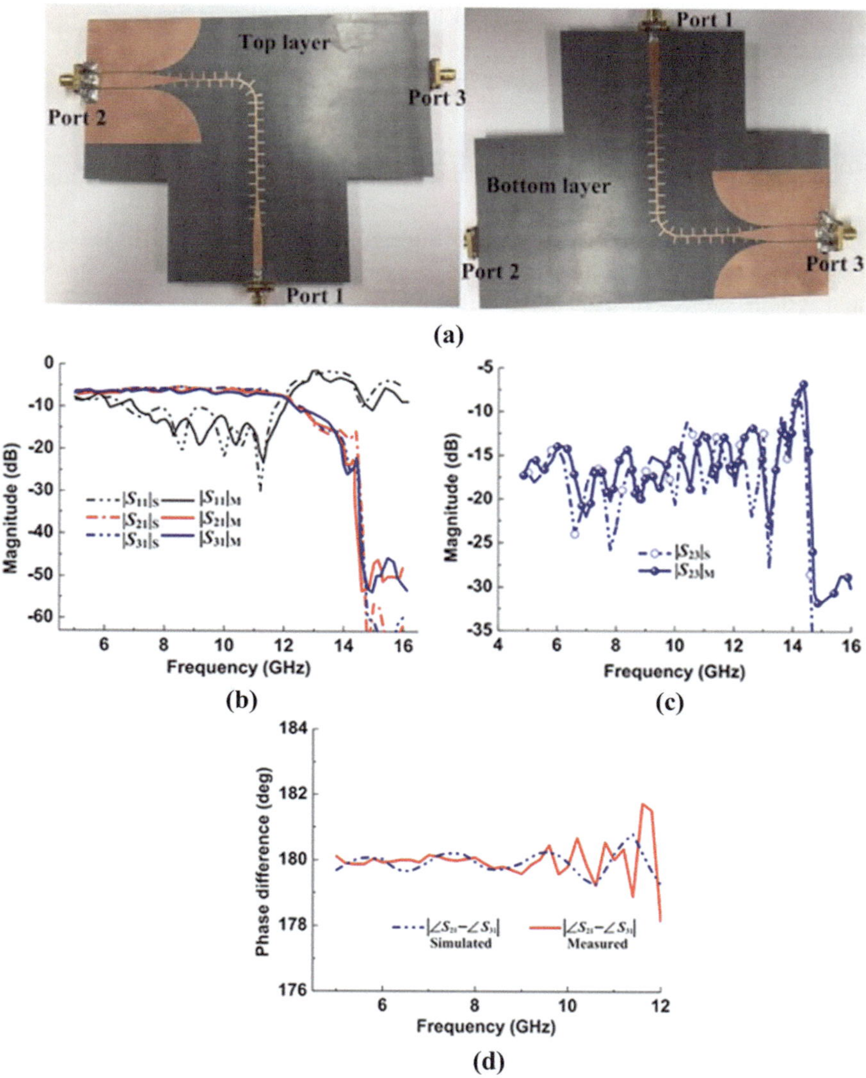

Fig. 7.6 **a** Photograph of the fabricated prototype of the wideband power divider. Measured and simulated results of the wideband power divider. **b** Transmission and insertion loss of the power divider. **c** Power divison isolation and **d** Phase difference [11]

SSPPs. The wideband power divider structure has a gradual transition from an H-shaped spoof surface plasmon polaritons (SSPPs) transmission line (TL) to a pair of U-shaped SSPPs TLs on the ultrathin Polyimide substrate, providing excellent flexibility and high isolation. Between the two U-shaped SSPPs TLs, a fragment of graphene flake is loaded as an isolation resistive film [15]. The 6.8–11.2 GHz band is covered by the flexible broadband power divider's high isolation, good matching, and equal power dividing with minimal loss when operating in flat, severely bent, and twisted conditions.

7.3 Phase Shifter Using Spoof Surface Plasmon Polariton

The phase shifter is a very important component in microwave, Millimeter wave, or Terahertz systems, such as phased arrays, beam forming networks, and measurement systems. Phase shifters with different performance are developed, such as wideband phase shifters and ultra-flat phase shifters. Researchers are drawn to the terahertz gap because of its intriguing qualities, which include the possibility of using it for high-resolution imaging and ultra-wideband communication, both of which call for reliable and effective terahertz circuits and components. The ideal substitute for implementing terahertz integrated circuits is terahertz spoof surface plasmon polaritons.

An unprecedented insertion phase versus loss performance is presented by the terahertz spoof surface plasmon polariton delay lines for terahertz phase shifters. The suggested delay lines' electrical lengths can be adjusted simply by varying the corrugation depths, giving the designer the ability to maintain a constant delay line's physical length. Four distinct delay lines, each measuring 1052.5 μm in length, are designed and their electrical lengths are chosen to correspond to the four distinct phase states of a two-bit phase shifter. Insertion and return losses on average of −2.3 and −18 dB, respectively, and an extremely low average phase error of 0.6% an unprecedented 90°/dB figure-of-merit at 0.3 THz is achieved in comparison to all other types of planar waveguides [16]. The suggested delay lines prove to be a very promising and high-performance option for all the waveguide-based parts that are required for the terahertz integrated circuits.

Their strong field confinement ability at low frequencies has caused widespread concern regarding SSPP. The need for nonreciprocal spoof plasmonic platforms with diode functionalities is critical for miniaturized integrated circuits. By building a plasmonic coupled line by flip-stacking two corrugated metallic strips, one can increase the mode coupling between evanescent waves with opposing transverse spin angular momenta, thereby realizing nonreciprocal phase shifting in SSPPs while utilizing the transverse Faraday effect. Two circularly polarized SSPP modes with distinct propagation constants across a wide bandwidth are created by transversely magnetized ferrite cladding. Figure 7.7a shows the schematics of the nonreciprocal SSPP phase shifter. Red and blue arrows show forward and backward propagating SSPPs, respectively. The bias magnetic field is applied along the y-direction. Inset is the structure of the periodic corrugated metallic strips. Figures 7.7b, c shows the switching of near-field chirality in forward and backward SSPPs [17]. To confirm that time-reversal symmetry in SSPPs is broken, a nonreciprocal phase shifter compatible with standard microstrip is designed. A differential phase shift up to 46.2°/cm from 12 to 15 GHz is shown by microwave measurement. This approach provides a different

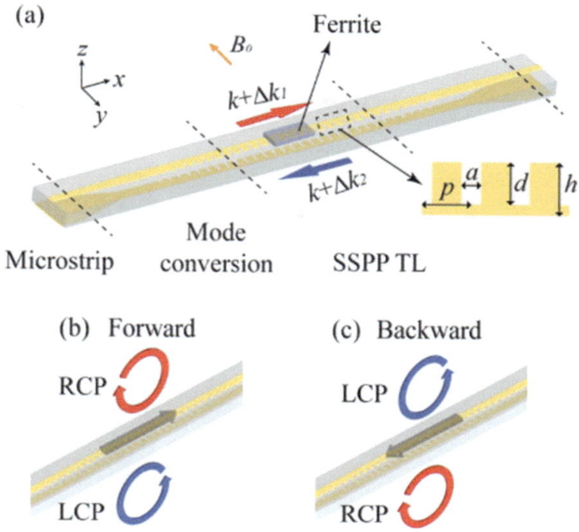

Fig. 7.7 Schematics of the nonreciprocal SSPP phase shifter. **a** The designed structure. Red and blue arrows show forward and backward propagating SSPPs, respectively. The bias magnetic field is applied along the y-direction. Inset is the structure of the periodic corrugated metallic strips. **b**, **c** Switching of near-field chirality in forward and backward SSPPs [17]

route for nonreciprocal spoof interconnects because of the benefits of contactless ferrite integration and strong field confinement.

Phase shifter application using SSPP is a 180° directional coupler made up of a 90° phase shifter and a quadrature hybrid. Figure 7.8 shows the photographs of the fabricated samples; (a) Top and (b) Bottom views of the fabricated SSPP mono-pulse antenna and microstrip mono-pulse antenna samples [18]. The desired characteristic impedance and electric length of the quadrature hybrid branch lines and phase shifter unit structure can be approximately realized by adjusting the unit length and stub length. The design requirement of a quadrature hybrid is satisfied by the horizontal and vertical branch lines with the equivalent phase shift of 90°, which are intended to have the equivalent characteristic impedance in ohms of 35.4 and 50, respectively. Four SSPP units are used in the design of the phase shifter with 50 Ω line, and each SSPP unit can produce an equivalent electrical length of 66.5°. In actuality, the phase shifter's constant length allows for the realization of various phase shift values, which is convenient for feeding network design. In the 9.5–10.5 GHz band, isolation($|S_{21}|$) is better than 17 dB, and reflection (S_{11}) is less than −17.6 dB when port 1 is used as the input port. With a bandwidth of 9.5–10.1 GHz, the output balance is 0.5 dB. When port 2 is utilized as the input port, the output balance of 0.5 dB falls within the bandwidth of 9.5–10.1 GHz, and reflection (S_{22}) is less than −18.9 dB in the band from 9.5 to 10.5 GHz. In the 9.64–10.2 GHz band, 0° ± 5° phase difference can be attained when port 1 serves as the input port. In the 9.82–10.3 GHz band, a 180° ± 5° phase difference can be reached when port 2 is used as the input port.

The tunable phase shifter is greatly needed due to its powerful control of electromagnetic waves. Tunable phase shifters based on different principles have been explored in this section. It is possible to investigate a tunable 360° phase shifter based on an active spoof surface plasmon polaritons (SSPP) waveguide. The device

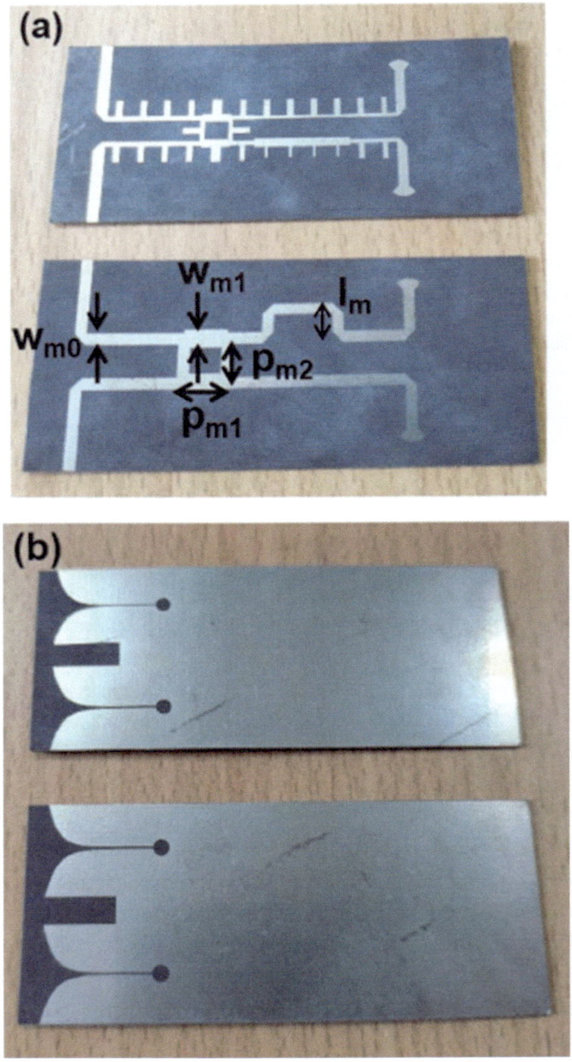

Fig. 7.8 Photographs of the fabricated samples. **a** Top and **b** bottom views of the fabricated SSPP mono-pulse antenna and microstrip mono-pulse antenna samples [18]. Reproduced with permission from the IEEE

is made up of seven varactor diodes applied to a passive SSPP waveguide [19]. The tunable phase shifter operates based on the dispersion curve being tunable with varying applied capacitance, allowing for the achievement of different transmission phases for a single unit cell with varying capacitances; a tunable phase variation of 180–360° within 6.6–7.2 GHz is achieved. The widespread applications of traditional phase shifters (PSs) are limited because they primarily concentrate on phase performance and infrequently consider the phase shift between different electromagnetic (EM) modes. The time delay of signals arises from the strong field confinement of EM waves supported by spoof surface plasmon polaritons (SSPPs). By adjusting the plasmonic structure parameters, SSPPs can be engineered in any way due to their flexible dispersion behaviors. Here, we suggest a multimode, reconfigurable SSPP PS that is loaded with varactor diodes, taking advantage of these qualities. Phase differences between various SSPP modes can be realized and the phase shift of each mode can be reconfigured by manipulating the dispersion behaviors.

Figure 7.9a shows the schematic of the SSPP PS with the bottom metal ground. Figures 7.9b, c show the transmission coefficients at different capacitances. Figure 7.9d, e show the reflection coefficients at different capacitances. Figure 7.9f–h shows the phase differences between the fundamental modes, high-order modes, and

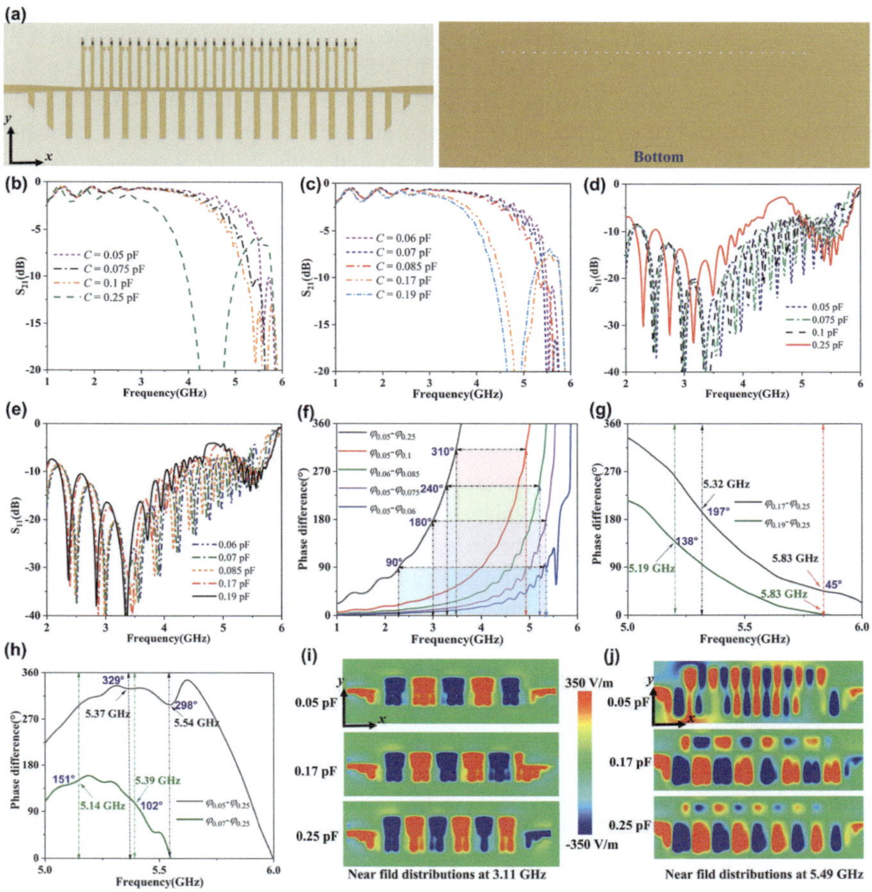

Fig. 7.9 Simulated results of the SSPP PS at different modes. **a** Schematic of the SSPP PS with the bottom metal ground; **b** and **c** Transmission coefficients at different capacitances; **d** and **e** Reflection coefficients at different capacitances; **f–h** Phase differences between the fundamental modes, high-order modes, and fundamental and high-order modes, respectively; **i, j** Near-field distributions at 3.11 and 5.49 GHz as the capacitances are 0.05, 0.17, and 0.25 pF, respectively, [20]. Reproduced with permission from the IEEE

fundamental and high-order modes, respectively, whereas (i), (j) show the near-field distributions at 3.11 and 5.49 GHz as the capacitances are 0.05, 0.17, and 0.25 pF, respectively, [20]. According to experimental results, by varying the bias voltages applied to the varactors, the proposed SSPP.

PS can achieve phase shifts of approximately 240°, 241°, and 340° among the fundamental (even) modes, the high-order (odd) modes, and the fundamental and high-order modes, respectively. Furthermore, the suggested SSPP PS can continuously adjust each mode's phase difference. There is good consistency between measurements and simulations. This work has a promising application in Terahertz circuits and lays the foundation for SSPP applications in large-scale integrated circuits and wireless communication systems.

Filtering phase shifters that can be adjusted based on liquid crystal materials and spoof surface plasmon polaritons (SSPP) offer both continuously tunable phase shifts and self-embedded fixed filtering performance [21]. Transmission parameter representations and equivalent circuit analysis are used to explain the fixed stopband caused by split ring resonators (SRRs). Additionally, the SSPP waveguide's slow wave characteristics allow for the intrinsic realization of a higher tuning rate and figure of merit (FoM), indicating the possibility of miniaturization and a reduction in insertion loss for particular phase shifts.

7.4 Conclusion

In this chapter, we studied different types of power dividers and phase shifters which form basic building blocks for different wireless communication systems may it be microwave, millimeter wave, or terahertz communication. We also studied different types of power dividers such as beam splitters. Different types of phase shifters using spoof surface plasmon polariton are also reported here. The reconfigurability principle is also discussed in this chapter.

References

1. Aziz, A., Fan, Y., He, P.H., Zhang, H.C., Ali, I., Tang, W.X., Cui, T.J.: Spoof surface plasmon polariton beam splitters integrated with broadband rejection filtering function. J. Phys. D, Appl. Phys. **54**(33), 335105 (2021)
2. Li, M., Wu, Y., Qu, M., Li, Q., Liu, Y. A novel power divider with ultra-wideband harmonics suppression based on double-sided parallel spoof surface plasmon polaritons transmission line. Int. J. RF Microw. Comput. Aided Eng. (2018)
3. Xiao, B., Kong, S., Chen, J., et al.: A microwave power divider based on spoof surface plasmon polaritons. Opt. Quant. Electron. **48**, 179 (2016)
4. Ye, L., et al.: High-performance spoof surface plasmon polariton waveguides and splitters based on Greek-cross fractal units. J. Phys. D Appl. Phys. **53**(23), 235502 (2020)
5. Zhou, S., Lin, J.Y., Wong, S.W., et al.: Spoof surface plasmon polaritons power divider with large isolation. Sci. Rep. **8**, 5947 (2018)

6. Pan, B.C., Yu, P., Liao, Z., Zhu, F., Luo, G.Q.: A compact filtering power divider based on spoof surface plasmon polaritons and substrate integrated waveguide. IEEE Microw. Wirel. Compon. Lett. **32**(2), 101–104 (2022)
7. Wu, Y., Li, M., Yan, G., Deng, L., Liu, Y., Ghassemlooy, Z.: Single-conductor co-planar quasi-symmetry unequal power divider based on spoof surface plasmon polaritons of bow-tie cells. AIP Adv. (2016)
8. Farokhipour, E., Komjani, N., Chaychizadeh, M.A.: An ultra-wideband three-way power divider based on spoof surface plasmon polaritons. J. Appl. Phys. **124**(23), 235310 (2018)
9. Zhang, X., Tang, W.X., Zhang, H.C., Xu, J., Bai, G.D., Liu, J.F., et al.: A spoof surface plasmon transmission line loaded with varactors and short-circuit stubs and its application in Wilkinson power dividers. Adv. Mater. Technol. **3**(6), 1800046 (2018)
10. Moznebi, A.-R., Afrooz, K.: Bandwidth enhancement of a half mode substrate integrated waveguide filtering power divider using spoof surface plasmon polariton. J. Phys. D Appl. Phys. **55**(2), 025103 (2021)
11. Feng, W., Che, W.: Wideband filtering power dividers using single- and double-layer periodic spoof surface plasmon polaritons. Int. J. RF Microw. Comput. Aided Eng. (2019)
12. Aziz, A.: A compact broadband power divider based on spoof surface plasmon polaritons. Laser Phys. **30**(1), 016207 (2019)
13. Li, J., Lv, S., Wu, S., Li, Q., Xu, K.-D.: Millimeter-wave dual-wideband filtering power divider using groove gap waveguide and spoof surface plasmon polariton. IEEE Trans. Compon. Packag. Manuf. Technol. **14**(6), 1089–1097 (2024)
14. He, P.H., et al.: Analysis, reduction, and utilization of loss in reconfigurable spoof surface plasmon polaritons. IEEE Trans. Microw. Theory Tech. **71**(3), 945–955 (2023)
15. Wu, B., Zu, H., Xue, B., Zhao, Y., Cheng, Q.S.: Flexible wideband power divider with high isolation incorporating spoof surface plasmon polaritons transition with graphene flake. Appl. Phys. Exp. **12**(2) (2019)
16. Unutmaz, M.A., Unlu, M.: Spoof surface plasmon polariton delay lines for terahertz phase shifters. J. Lightwave Technol. **39**, 3187–3192 (2021)
17. Yan, H., Jing, L., Zhao, J., Niu, C., Zhang, Y., Liuge, D., Wang, Z.: Broadband nonreciprocal spoof plasmonic phase shifter based on transverse Faraday effects. Opt. Express **30**, 24000–24008 (2022)
18. He, P.H., Ren, Y., Shao, C., Zhang, H.C., Zhang, L.P., Cui, T.J.: Suppressing high-power microwave pulses using spoof surface plasmon polariton mono-pulse antenna. IEEE Trans. Antennas Propag. **69**(12), 8069–8079 (2021)
19. Tian, D., Kianinejad, A., Zhang, A.: Tunable 360 degrees phase shifter based on spoof surface plasmon polaritons waveguide. In: 2019 Photonics & Electromagnetics Research Symposium—Fall (PIERS—Fall), Xiamen, China, pp. 182–184 (2019)
20. Gao, X., et al.: Multimode and reconfigurable phase shifter of spoof surface plasmons. IEEE Trans. Antennas Propag. **71**(6), 5361–5369 (2023)
21. Ding, C., Meng, F.-Y., Han, J. -Q., Mu, H.-L., Fang, Q.-Y., Wu, Q.: Design of filtering tunable liquid crystal phase shifter based on spoof surface plasmon polaritons in PCB technology. In: IEEE Transactions on Components, Packaging and Manufacturing Technology, vol. 9, no. 12, pp. 2418–2426 (2019)

Chapter 8
Spoof Surface Plasmon Polariton for Biomedical Applications

8.1 Introduction

The precise and highly sensitive determination of the water content of human biological tissues is necessary to meet various demands and proposals in the fields of precision medicine, early diagnostics, and point-of-care testing (POCT). In vivo assessment of the human stratum corneum's water content for cosmetic purposes and the diagnosis of superficial tumor are two examples of fields that are encouraged to grow as a result. To accomplish these objectives, we must measure the water content and dielectric characteristics of human skin tissues to accurately identify between normal and malignant tissue types.

An analytical tool that combines a biological component and a physicochemical detector to detect chemicals is called a biosensor. Resonantly enhanced dielectric sensing exhibits superior sensitivity and accuracy based on the signals measured from relative resonance shifts that are immune to signal fluctuations. These sensing resonators find applications in different biosensing applications. Although label-free optical biosensors operating in the visible range are commonly utilized, their sensitivity to water is limited, necessitating their combination with other mechanical techniques to determine the water content. Recent developments in microwave, millimeter, and terahertz wave sensing techniques have been attributed to their highly sensitive, dependable, and accurate bio-detection capabilities, which enable label-free, non-contact, and non-destructive inspection of chemical and biological materials.

By measuring the response of the molecules after excitation, microwaves can determine the composition of specific tissues and fluids by interacting with the materials based on their molecular structures. Thus, before notable alterations in biological tissues are noticed, examining the microwave signals derived from non-biological parameters may disclose a variety of early disease indicators. The terahertz frequency band in the electromagnetic spectrum is located between the microwave and far-infrared regions. THz waves are useful for creating extremely sensitive, label-free,

S. K. Koul and S. Sharma, *Microwave Integrated Circuits Based on Spoof Surface Plasmon Polariton*, Signals and Communication Technology,
https://doi.org/10.1007/978-981-96-9041-1_8

and non-intrusive sensors because they are non-ionizing and have wavelengths that are similar to the sizes of most biomolecules. Nevertheless, because aqueous samples absorb EM waves due to their high-water content and are highly lossy, THz waves for sensing are only useful for dry samples.

The field of biological sensing, in particular bio-liquids like aqueous solutions of proteins, DNA, and cells is expected to be greatly impacted by THz sensing. The relatively long wavelength of THz waves about the size of the sample that needs to be characterized makes them unsuitable for use in these applications. Consequently, methods that improve light-matter interactions in the THz regime by tightly confining the electromagnetic field are very desirable. One way to do this is to use metamaterials, which are made up of strongly enhanced localized electromagnetic fields made of periodic metallic or dielectric resonators.

For biomedical applications like sensing and imaging, the unique properties of spoof surface plasmon modes makes it a strong contender. A highly sensitive surface to any change in the surrounding dielectric environment of the sensing device is provided by its properties, such as strongly confining electromagnetic fields at the metal surface boundary. Spoof surface plasmon polaritons (SSPP) have controllable dispersive properties that enable one of the most efficient and systematic manipulations of electromagnetic waves. In addition to being fascinating from physics standpoint, their unique characteristics such as the slow light close to the frequency cut-off—allow for a variety of applications.

8.2 Spoof Surface Plasmon Polariton for Biomedical Applications

The traditional approach of measurement not only results in large radiation losses but also is unable to ensure a repeatable measurement free from uncertainty caused by surrounding electromagnetic interference. There is just not enough confinement in the propagating energy field to produce a highly energetic beam that is not deflected by the surrounding electromagnetic disturbance. The shifts in the resonance frequencies that determine the glucose level detection are challenging to achieve in the absence of a shield against ambient electromagnetic distraction. This explains why spoof surface plasmon-based topology is becoming more and more common.

8.3 Cancer Detection

Spoof SPPs have recently gained great interest in biosensing due to the strong enhancement of the localized electric field. For the diagnosis of ovarian cancer, a sensitive spoof SPP biosensor equipped with split ring resonators (SRRs) can be employed. In place of the traditional periodic grooves, the SRRs are spaced in series on a metal line to produce a sharp resonance and preserve the cut-off frequency. Figure 8.1a shows the sliced samples on a glass slide whereas Fig. 8.1b shows the fabricated biosensor attached to a sliced sample [1]. The suggested device is extremely sensitive for detecting permittivity differences because of this resonance, which multiplies the localized field by a factor of 250. Figure 8.2a shows a biosensor with a liquid wall for solution detection whereas Fig. 8.2b shows the measured resonant frequencies for sucrose solutions with various concentrations. The suggested biosensor successfully detects ovarian clear cell carcinoma (OCCC) and serous ovarian cancer (SOC) tissues, and it has the advantage of high sensitivity. The biosensor's intrinsic resonance frequency is 53.990 GHz; with normal, SOC, and OCCC tissues, it shifts to 53.814–53.968, 53.698–53.872, and 53.719–53.845 GHz, respectively. A reasonable reference for cancer detection is the average point of all ovarian tissues, which is at 53.812 GHz, which is much below the normal tissues. Additionally, all cancerous tissues have a lower average transmittance than SOC tissues; this difference can be used to distinguish between SOC and OCCC tissues. As a result, the spoof SPP biosensor offers an efficient, accurate, and label-free method of ovarian cancer detection.

The confinement and enhancement of microwave by SSPPs allows a high density of independent subwavelength sensor elements to be constructed in millimeter-sized arrays. The biosensor can be used to distinguish biological tissues because different water contents will cause different microwave attenuations. A microwave biosensor, based on the spoof surface plasmon polaritons (SSPPs) excited by a planar plasmonic waveguide, which can measure in vivo the water content of human skin tissues timely and accurately [2]. The microwave biosensor is designed with the planar periodic slow wave structured unit cell; it is a biosensor with a human biological tissues model based on the Debye model. A SSPPs biosensor is verified with the vector network analyses using a human finger as a real model. Experimental results of the biosensor with/without normal human finger skin tissues and pure water have good agreement with the simulation. The SSPPs biosensor has potential applications for the areas

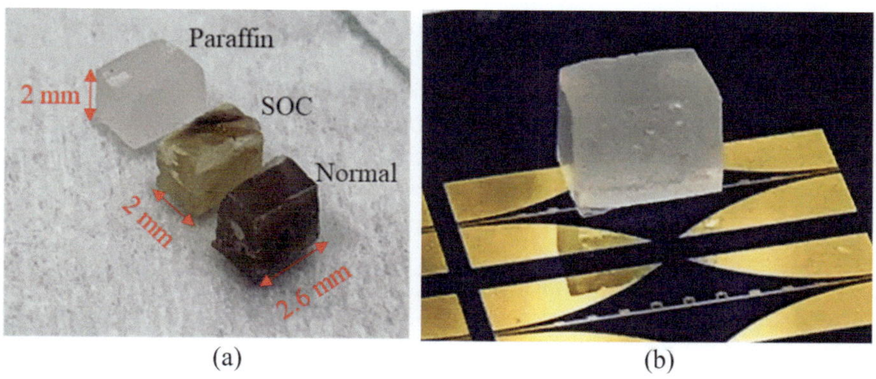

Fig. 8.1 **a** Sliced samples on a glass slide **b** the fabricated biosensor attached with a sliced sample [1]

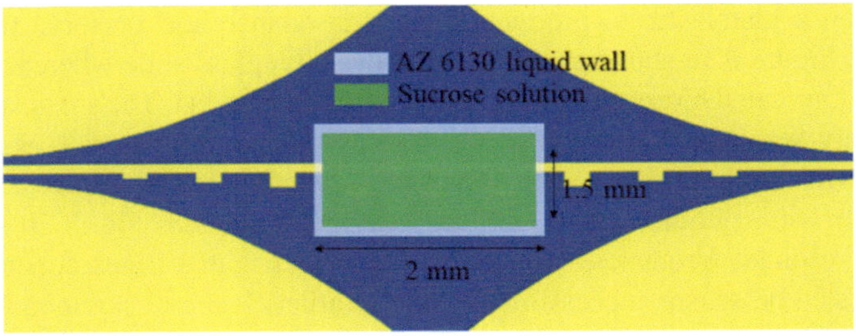

(a)

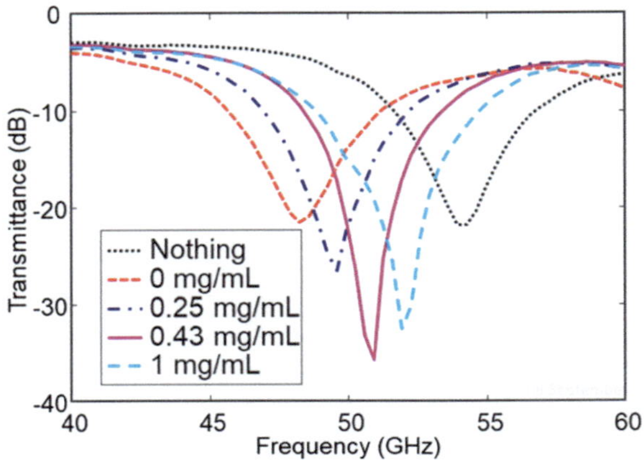

(b)

Fig. 8.2 **a** Biosensor with a liquid wall for solution detection. **b** Measured resonant frequencies for sucrose solutions with various concentrations [1]

of in vivo measurement of the human stratum corneum water content with topical cosmetics, early diagnostics of superficial tumors such as malignant melanoma and lymphatic cancer, and in guiding tumor resection. To enhance the detection sensitivity and accuracy of the water content in human biological tissues, localized spoof surface plasmon polaritons (SSPPs) excited by a periodic metal–dielectric planar plasmonic slow waveguide are utilized. It is relatively straightforward to integrate those sensors into microfluidics chips, making plasmonic structures promising for use in next-generation modern biosensors. This microwave biosensor has potential applications for routine point-of-care testing (POCT), clinical evaluation, real-time diagnosis of tumor.

The structure of spoof surface plasmon polaritons (SSPPs) can be used as a biosensing sensor in the THz range. For biosensing, sensor amplitude and resonance accuracy are critical factors. The resonance position and amplitude of SSPPs are determined by their momentum matching and gap distance, respectively. The sample is placed between the prism base and the SSPP structure to facilitate biomolecular sensing [3]. The resonance position has a significant error, and the momentum matching condition in the current study ignores the impact of the sample refractive index. The momentum matching condition has been corrected to account for the sample refractive index. In the meantime, various types of brain lesions can be detected in the frequency range as an application of biosensors, according to a comparative study of surface plasmon resonance (SPR) sensing performance based on frequency and angle variations. The sensing sensitivity for the frequency region is superior to that of the angle region.

Nanowire metamaterials are essential for many different fields of application, such as clinical cancer detection and antenna research. Examining surface wave characteristics at the ellipsoidal nanowire metamaterial boundary, the effective properties of the ellipsoidal medium are ascertained [4]. Because the selected model shares characteristics with actual biological systems that contain cancer cells, such as ellipsoidal inclusions, the former may lead to a potential application in clinical practice dealing with cancer detection, particularly by providing a deep insight into the derived medium treatment approach. Prospects may involve utilizing effective medium approximation models to translate the obtained effective medium properties into clinical practice for cancer detection.

8.4 Tumor Detection

It is possible to realize a highly sensitive microfluidic sensing method based on spoof surface plasmon polaritons (SPPs) for the terahertz (THz) region of the electromagnetic spectrum [5]. We enable dielectric sensing of liquids by utilizing these highly confined electromagnetic modes by integrating a microfluidic channel in a spoof SPP waveguide. A domino waveguide, or a set of rectangular metal blocks arranged periodically on top of a metal surface, facilitates the propagation of spoof SPPs. Figure 8.3 shows the schematics of the homogeneous (a) and the tapered (b) spoof SPP domino waveguides, with lateral width W, period p, groove depth h, and groove gap width g as indicated in the inset figure. A micro-capillary is positioned in the center groove. The center of the capillary is aligned with the top of the domino blocks. The refractive index of a liquid passing through a microfluidic channel that crosses the waveguide to produce an interaction volume on the nanolitre scale has a significant impact on the transmission of spoof SPPs along the waveguide. Moreover, a tapered waveguide can accomplish additional electromagnetic field confinement by utilizing the insensitivity of the fundamental spoof SPP mode of the domino waveguide to the lateral width of the metal blocks. This method is designed to detect individual subwavelength microparticles flowing in the liquid with a high degree of sensitivity. Figure 8.4 shows the sensing in a single micro-particle passing through the capillary; (a) Transmission spectra of a homogeneous domino waveguide with a single spherical particle in a liquid inside the capillary, (b) Transmission spectra for the tapered domino waveguide with capillary and particle, with particle parameters being the same as in (a). The development of spoof SPP-based THz lab-on-a-chip

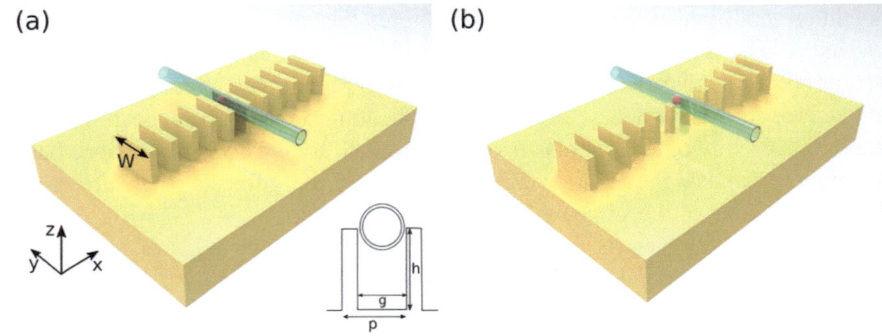

Fig. 8.3 Schematics of the homogeneous (**a**) and the tapered (**b**) spoof SPP domino waveguides [5]

microfluidic devices that can analyze biological liquids like proteins and circulating tumor cells in buffer solution appears to be aided by these promising results.

8.5 Biomedical Imaging

One potential low-cost near-field imaging system is a spoof surface plasmon (SPP) transmission line (TL) system. Investigating wearable medical imaging equipment, such as that used to detect blood or skin cancer, may be feasible. The concept could lead to 2-D scanning even though the reported structure was designed for 1-D scanning [6]. These SPPTLs can be constructed in a parallel array, or a single long SPPTL can be bent in a meander line to span a two-dimensional space. An accurate image of a sparse scene can be obtained through signal processing with a compressive sensing algorithm. A single-radiator system needs a highly dispersive element, in contrast to traditional multiple antenna-based imaging. Frequency diversity can be used to replace spatial diversity in conventional systems by utilizing the dispersion of radiated pattern. Experiments are conducted on three SPP TLs with varying groove depths to test the concept. It is discovered that a dispersive SPP TL operating in the 1–5 GHz frequency range can achieve a high subwavelength resolution of 5 mm (~0.05 λ, where λ is the air wavelength at 3 GHz). It was discovered that the propagation constant is directly related to imaging resolution; therefore, a higher effective refractive index of TL can result in a higher resolution.

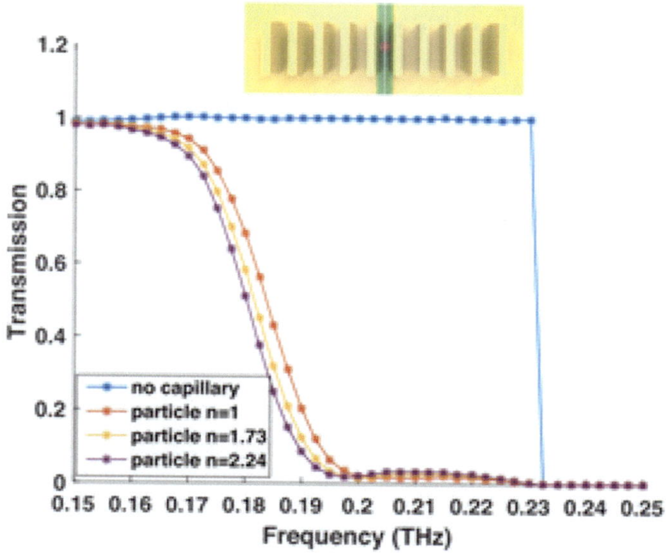

(a)

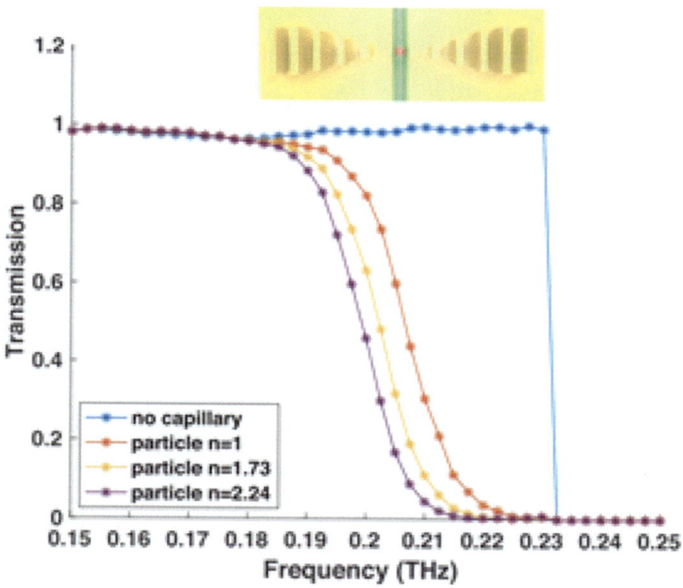

(b)

Fig. 8.4 Sensing a single micro-particle passing through the capillary. **a** Transmission spectra of a homogeneous domino waveguide with a single spherical particle in a liquid inside the capillary. **b** Transmission spectra for the tapered domino waveguide with capillary and particle, with particle parameters being the same as in (**a**) [5]

8.6 Biosensing Applications

The biosensing application of SSPP-based resonator helps us to sense a few hundreds of bacteria. A highly sensitive, non-intrusive, label-free THz biosensor with a microfluidic approach can be used to detect the concentration of glucose in an aqueous medium. Since the THz waves are currently propagating on the chip, a highly sensitive measurement can be made because the electromagnetic environment surrounding the confined wave is clearly defined. Using resonators is a common way to increase the effect. Despite the widespread use of metamaterials, the method of creating whispering-gallery mode (WGM) micro resonators directly from the SPP waveguide is investigated. The sensitivity of WGM micro resonators is higher than that because they have substantially higher quality factors.

The WGM resonator produces multiple resonances (>10) between 0.50 and 0.54 THz by creating a frequency comb at the same frequency range as the waveguides [7]. Similar to WGM resonators, the optimal distance between waveguides and resonators needs to be determined to attain the highest possible quality factor. Figure 8.5 shows schematic of the SPP resonator used for biomedical sensing. The small coupling distances between the waveguide and resonator, which are on the order of 1/100th of a wavelength, are caused by the strongly confined propagation on the SPP waveguide. Figure 8.6 shows the intensity of resonance which depends on the number of bacteria on the surface.

The quality factors of an SPP WGM resonator can reach about 500. Antibiotic-resistant bacteria S. Aureaus is the biomedical target. The diameter of S. Aureaus cells ranges from 0.7 to 1.5 μm. Bacteria can be concentrated on the resonator surface through surface functionalization, which binds them to gold surfaces. Using a material with well-known electromagnetic properties has an added benefit in that it makes simulation and experiment comparisons more accurate. To replicate the realistic distribution on a functionalized surface, randomized particle distributions were employed. The sensor is an appropriate concept for biosensing because it is highly sensitive to even minute concentrations of a few hundred bacteria.

An extremely effective surface plasmon polariton (SSPP) end-fire spoof sensor for on-body glucose monitoring and monitoring of aqueous glucose solutions [8]. To suppress the sidelobes, which are ineffective in glucose sensing, two triangular ground planes were added to the sensor's coplanar waveguide (CPW) port. The suggested SSPP end-fire sensor's slow wave characteristic has further benefited the measurement of glucose concentrations by enhancing sensitivity. With a group velocity like the phase velocity of light propagating in a glucose solution, the guiding body of the sensor was designed to mimic an electromagnetic beam propagating in a dense dielectric medium. According to the experimental results of this work, which are in line with our analysis, the slow electromagnetic wave gradually transitioning from the SSPP end-fire sensor to the glucose water solution has successfully made it possible to measure and display the concentrations of glucose in a solution as highly observable resonant frequency shifts in S_{11} parameters. During on-body measurements, the design has reached a high sensitivity of 150 MHz/mgmL^{-1}.

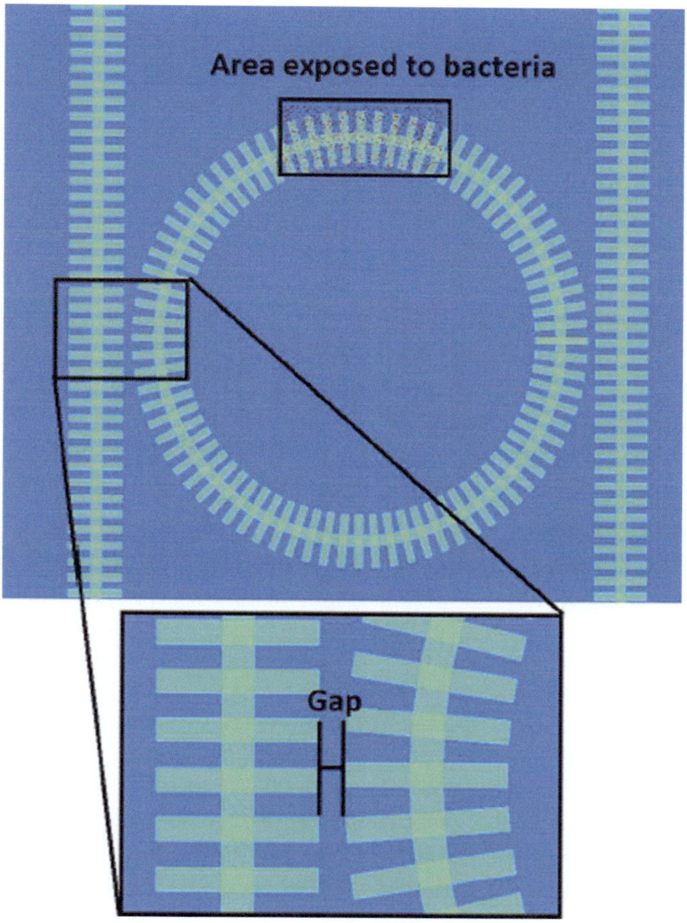

Fig. 8.5 Schematic of the SPP resonator used for biomedical sensing [7]

To achieve high sensitivity for even the smallest change in the dielectric value near its surface, the THz sensor unit makes use of the highly confined feature of the localized spoof surface plasmon (LSSP) resonator. Utilizing the LSSP resonator's reflection behavior (S_{11}) as the sensing response, the sensor is designed to operate at 1 THz. With a high-quality factor of 192, the sensor's design yielded a high sensitivity of 13.5 MHz/mgml^{-1}. At microwave frequencies, the sensor responds very sensitively, with a sensitivity of 1.2771e–4 MHz/mgml^{-1}. To obtain the sensor response in terms of DC voltage and to offer a proof of concept for the low-cost point-of-care (PoC) detection solution, a specially designed and developed read-out circuitry is also designed and developed.

The sensor is an LSSP ring resonator, which localizes and confines the electromagnetic field. This leads to an increase in the EM wave's interaction with the sample under test (SUT) and high sensitivity [9]. However, to use the suggested

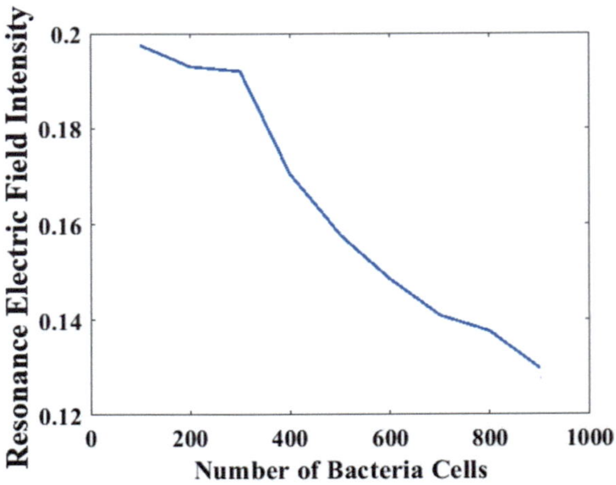

Fig. 8.6 Intensity of a resonance in dependence of the number of bacteria on the surface [7]

sensor in a more practical setting, a system with an inexpensive read-out circuitry that is simple to calibrate against various environmental factors like temperature, humidity, pressure, etc., as well as for any systematic errors present in the system, is needed.

A corrugated metal stripe with three rectangular grooves perforating it is the basis of a planar terahertz metamaterial sensor that is being studied. Ultrasensitive sensing can be realized and utilized because of the formation of Fabry–Perot resonance of the spoof surface plasmons mode on the corrugated metal stripe, which is associated with strong local field enhancement and high-quality factor [10]. The geometrical parameters of the unit cell are described in Fig. 8.7. Analyte materials and electromagnetic waves interact strongly, resulting in a figure of merit of 19.86 and frequency sensitivity of 1.966 THz per refractive index unit. In contrast, when the analyte thickness is less than 4 μm, this metamaterial sensor's film thickness sensitivity is greater than 52.5 GHz/μm. The sensor performance is significantly impacted by the thickness of the metal. These results provide a path toward the development of planar metamaterial structures into useful sensors in the terahertz range.

The frequency shift (FS) versus various analyte layer thicknesses is displayed in Fig. 8.8a. The inset of Fig. 8.8a makes it evident that the resonance frequency dip red shifts rapidly at first and stops at a maximum thickness of 11 μm as the analyte thickness increases. This suggests that increasing thickness does not affect the resonance frequencies. The dependence of the resonance frequency shift on the variation of the refractive index for the sensor with two distinct analyte thicknesses of $t = 2$ μm and $t = 11$ μm is displayed in the inset of Fig. 8.8b. Analyte thickness increases from $t = 2$μ m to $t = 11$ μm can result in a significant increase in sensitivity. Moreover, Fig. 8.8b presented a clear plot of the sensitivity versus the various analyte thicknesses. The analyte thickness affects the film thickness sensitivity, as shown in Fig. 8.8c. The film sensitivity can reach 52.5GHz/μm when the film thickness is 4 μm, and it increases roughly linearly when the film thickness is less than 4μm. Consequently, it is possible to realize ultrathin film thickness sensing. With the

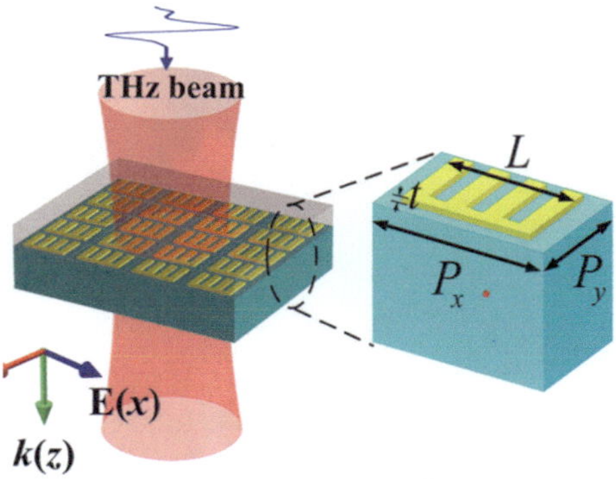

Fig. 8.7 Schematic diagram of the MMs structure covered by analyte layer [10]

analyte thickness fixed at 11 μm, Fig. 8.8d shows the variation of the Q factor and the corresponding FOM as a function of the analyte refractive index. The Q factors rise linearly from 57.46 to 87.32 when the refractive indices rise from 1.0 to 1.8. These extremely high Q values show improved detection accuracy and capability.

To identify skin abnormalities, point-of-care testing is crucial. To detect abnormalities in the skin, there must be enough depth and no damage [11]. To satisfy the needs of skin anomalies, a flexible microwave biosensor based on spoof surface plasmon polaritons was created. Figure 8.9 displays the microwave sensor that has been designed. The smaller sensing unit (0.6×0.5 mm^2) enables the sensor to have a higher resolution. The polyimide substrate of the sensor has a thickness of 0.1 mm. The sensor is flexible enough to be affixed to the skin for measurement, as Fig. 8.10 illustrates. Because of its flexibility, the sensor can easily measure any part of the human body, which has a few benefits for enhancing the sensor's convenience of detection. In this study, porcine skin was substituted for human skin as the experimental material. The S parameters of the sensor are displayed in Fig. 8.11 when it is loaded with MUT that has varying dielectric characteristics.

The resonant frequency shifts toward low frequency as permittivity rises. The differences in pork skin with varying heating is clearly visible, as seen in Fig. 8.12. The smooth surface of the unheated skin in Fig. 8.12 is burned gradually as the heating became deeper. The measurement results are displayed in Fig. 8.13. Four samples of pork skin with varying degrees of burns had various resonant frequencies and reflection coefficient magnitudes.

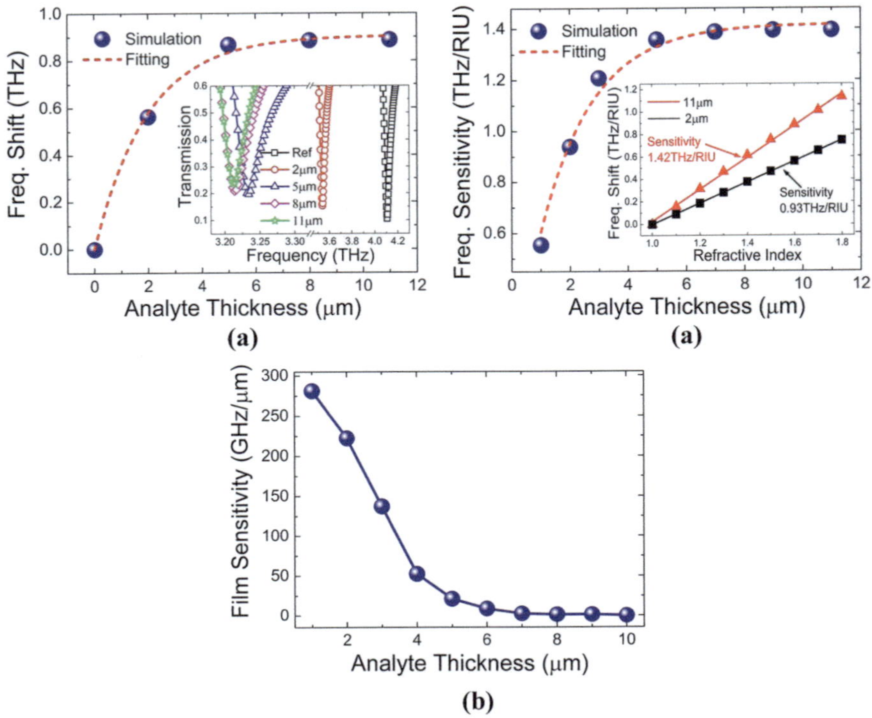

Fig. 8.8 **a** Frequency shift of the MMs structure with different analyte thicknesses; the inset shows the simulated transmission dips with different analyte thicknesses. **b** The frequency sensitivity versus different analyte thicknesses. **c** The film thickness sensitivity versus analyte thicknesses. **d** Variation of Q factor and FOM with different refractive indices of the analyte [10]

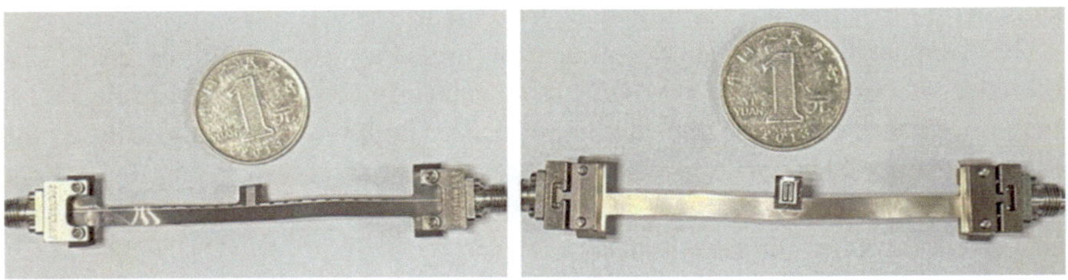

(a) (b)

Fig. 8.9 **a** Top view of sensor. **b** Bottom view of sensor [11]

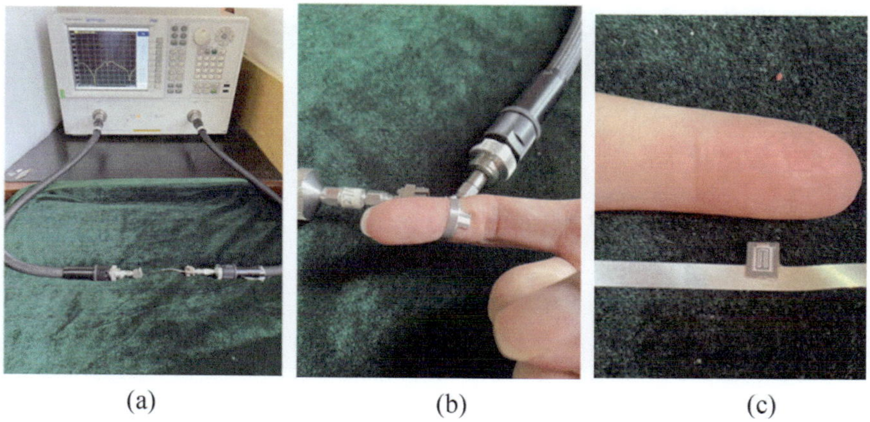

Fig. 8.10 **a** Measurement operation of sensor; **b** wrap the finger; **c** sensing unit of the sensor [11]

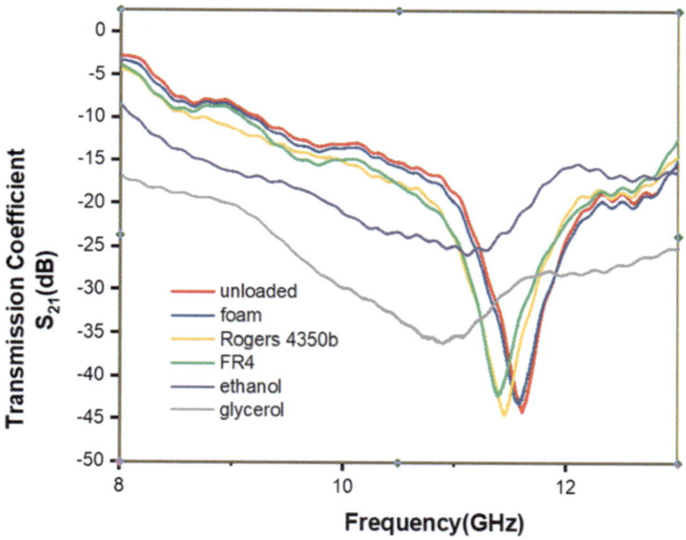

Fig. 8.11 Measurement results when sensor is loaded with MUT with different dielectric properties [11]

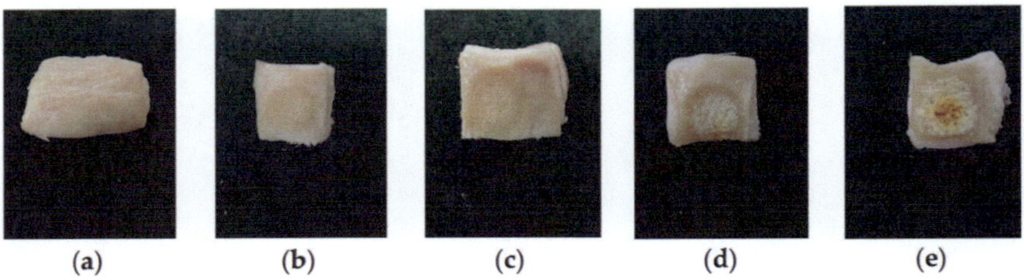

Fig. 8.12 Burns of different degrees: **a** burned for 0 s, **b** burned for 40 s, **c** burned for 80 s, **d** burned for 120 s, **e** burned for 160 s [11]

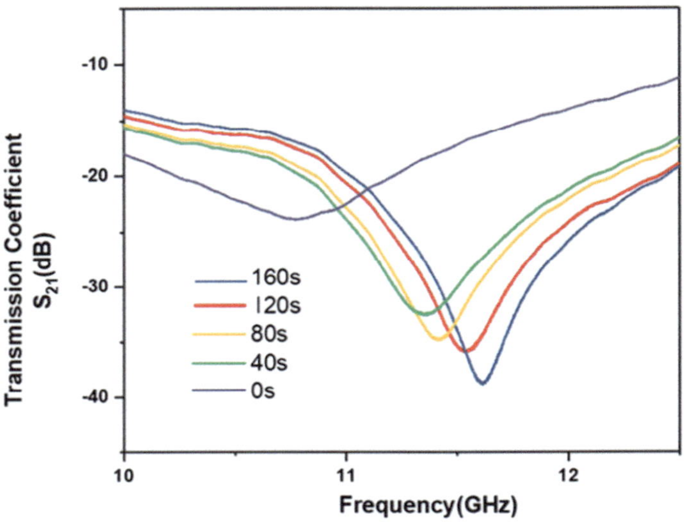

Fig. 8.13 Measurement results of burns of different degrees [11]

8.7 Conclusion

In this chapter, the advantages of using the spoof surface plasmon polariton and its applications in different biomedical domain such as sensing, imaging, tumor, and cancer detection are discussed. This chapter clearly provides the insight about how terahertz frequency is important for sensing application. It is also discussed that microwave frequency range is the appropriate solution for aqueous materials.

References

1. Zhang, Y., et al., Label-free diagnosis of ovarian cancer using spoof surface plasmon polariton resonant biosensor. Sens. Actuators B, Chem. **352** (2022). Art. no. 130996
2. Liu, G., Cheng, D., Zhang, B., Shu, G., Wang, J.: A microwave biosensor based on spoof surface plasmon polaritons for in vivo measurement of the water content of human skin tissues. J. Phys. D Appl. Phys. **52**(20) (2019)
3. Zhao, R., Lu, G., Yin, H., Liang, J., Zeng, D., Xiao, H.: Terahertz sensor study based on spoof surface plasmon polaritons. Int. J. Antennas Propag. **2504626**, 13 (2020)
4. Gric, T., Rafailov, E.U.: On the study of ellipsoidal nanowire metamaterials for biomedical applications. Opt. Quant. Electron. **55**, 993 (2023)
5. Ma, Z., Hanham, S.M., Huidobro, P.A., Gong, Y., Hong, M., Klein, N., Maier, S.A.: Terahertz particle-in-liquid sensing with spoof surface plasmon polariton waveguides. APL Photon. **2**(11), 116102 (2017)
6. Shi, Y., Rashid, A.K., Ma, D., Zhang, Q.: Spoof surface plasmon-based single-shot super-resolution compressive imaging. IEEE Trans. Plasma Sci. **48**(8), 2742–2750 (2020)
7. Klein, A.K., Bojahr, C., Stöhr, A.: Spoof plasmon polariton-antenna transitions for terahertz on-chip applications. In: 2021 15th European Conference on Antennas and Propagation (EuCAP), Dusseldorf, Germany (2021)
8. Kandwal, A., et al.: Surface plasmonic feature microwave sensor with highly confined fields for aqueous-glucose and blood-glucose measurements. IEEE Trans. Instrum. Meas. **70**, 1–9 (2021)
9. Pandit, N., Jaiswal, R.K., Pathak, N.P.: Towards development of a non-intrusive and label-free THz sensor for rapid detection of aqueous bio-samples using microfluidic approach. IEEE Trans. Biomed. Circuits Syst. **15**(1), 91–101 (2021)
10. Chen, X., Fan, W.: Ultrasensitive terahertz metamaterial sensor based on spoof surface plasmon. Sci. Rep. **7**, 2092 (2017)
11. Bai, J., Guo, H., Li, H., Zhou, C., Tang, H.: Flexible microwave biosensor for skin abnormality detection based on spoof surface plasmon polaritons. Micromachines **12**, 1550 (2021)

Chapter 9
Flexible or Textile Circuit Design Using Spoof Surface Plasmon Polariton

9.1 Introduction

The development of wearable sensors for applications in sports, recreation, health, and medicine has accelerated during the past ten years due to growing interest in the tattoo [1] or bandage-based Wireless Body Area Networks (WBANs). Figure 9.1 shows the demonstration of a tattoo antenna for on-body applications. These days, wearable and flexible electronics are very popular and are getting more and more attention from the researchers and commercial sector. The next generation of wearable electronics requires that wireless devices and antennas be worn directly on the human body surfaces [2]. This has motivated researchers to investigate a wide range of flexible substrates and how well they function on the body.

Flexible electronics and electronic textiles are an attractive replacement for the current electronics technology, which is based on rigid and brittle substrates [3, 4]. The advantages include their lightweight nature, energy efficiency, low manufacturing costs, reduced fabrication complexity, and the availability of affordable flexible films/substrates (such as paper, polyester, polyethylene, and plastics). One can also apply different flexible substrates directly to the body or integrate them with the clothing.

Some examples of these substrates are textile-based antennas, which use felt, silk, cotton, jean cloth, or any other appropriate textile combined with conductive elements like wires, fabrics, e-threads, and adhesive copper tape [5–7]. Surface waves known as spoof surface plasmon polaritons (SSPPs) occur on metal-dielectric interfaces and are becoming more and more well-known due to their qualities, which include low bending losses, high field confinement, and low crosstalk [8]. Because of its intrinsic qualities, SSPP is crucial to the design of textile or flexible circuits. SSPP is used in the design of many flexible components, including sensors, power dividers, transmission lines, antennas, and filters [9–11].

S. K. Koul and S. Sharma, *Microwave Integrated Circuits Based on Spoof Surface Plasmon Polariton*, Signals and Communication Technology,
https://doi.org/10.1007/978-981-96-9041-1_9

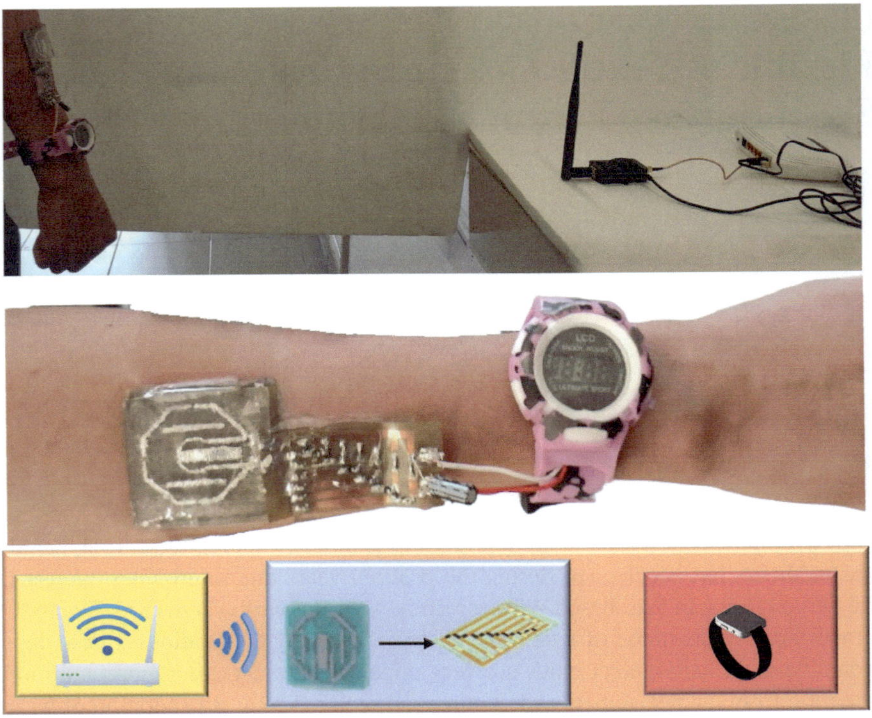

Fig. 9.1 Demonstration of tattoo antenna for on-body application [1]

9.2 Need for Textile or Flexible Circuits in Our Day-To-Day Life?

Textile or flexible circuits find increased application in wearable electronic devices due to technologies dedicated to health care, telemedicine, and entertainment applications. In health care, these flexible substrates worn on the body can closely monitor an individual's activities and enable multiple human physiological signals to be monitored in hospitals, residential, or work environments continuously [12–14]. Nowadays, healthcare systems are required to provide proactive wellness management and concentrate on the early detection and prevention of diseases.

Electronic textiles or e-textiles are fabrics that enable electronic components such as batteries, lights, sensors, and microcontrollers to be embedded in them. Many smart clothing, wearable technology, and wearable computing projects involve the use of e-textiles. Communication on the body degrades when no line-of-sight links are used and the body degrades the signal as human tissues provide high attenuation so by having a transmission system on textile the attenuation problem can be reduced and transmission can be improved [15, 16]. Figure 9.2 shows the transmission line on textile which can be used for on-body transmission so that attenuation can be

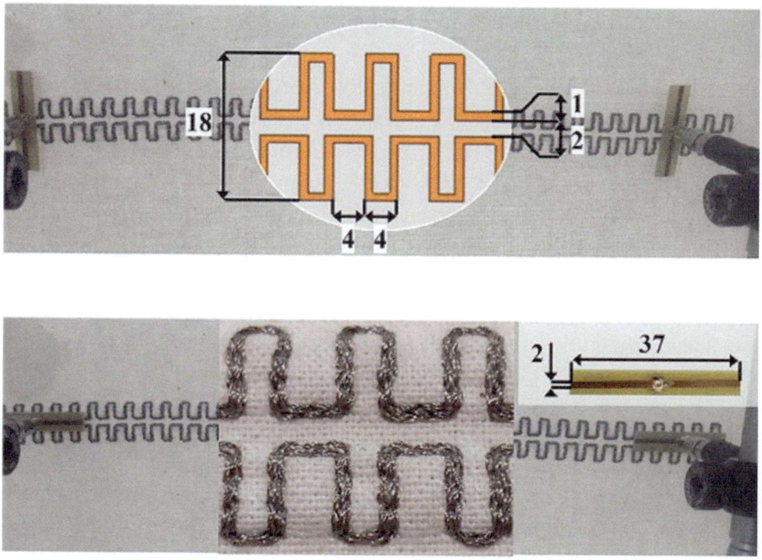

Fig. 9.2 Transmission line on textile fabric with stitching method [15]

reduced. At the endpoint, communication can be done by the antenna to reduce the need for connectors [17]. The use of an antenna can reduce the need for connectors and transmission can be done efficiently.

Due to this increasing demand for wearable technology and sensors to be integrated into the human body for sensing and health monitoring, several research studies are going on to develop materials or methods for wearable energy harvesters. Several works are going on to integrate RF energy harvesters into clothing that is worn on the human body. Due to their small thickness, small size, lightweight, and flexibility, flexible antennas can be used in miniaturized, portable, and foldable equipment. It has important research value in communication terminal equipment and satellites.

9.3 Advantage of Spoof Surface Plasmon Polariton Over Conventional Techniques

SSPP is a promising candidate as the spatial confinement limits the interaction of the guided wave with the human body and higher data security can be expected. Highly confined wave traveling in spoof surface plasmon makes it a perfect candidate for wearable application as coupling and radiation to near free space is almost negligible. Structural deformation study also plays an important role as fields are confined so structural deformation is less in SSPP. In [18], the structural deformation is studied and demonstrated that despite bending the transmission line, it has good transmission characteristics. The results are further compared with the results of bent microstrip line.

Figure 9.3a shows the SSPP Transmission line with different bending radii $R = 200$, 100, and 50 mm, and Fig. 9.3b shows the S-parameters of the transmission line with different bending radii; the results remain constant after considering bending effects. The use of SSPP and flexible technology makes previously impossible designs and technologies such as wearable and smart devices on the body possible. They provide strong capabilities for medical interventions, human–machine interfaces, and health monitoring.

monitoring. Clothing can be designed to create a highly effective and secure wireless body sensor by containing electromagnetic waves that are propagating around the human body by surface plasmon-like modes at radio communication frequencies.

Last ten years have seen the development of sensors, circuits, and displays. These devices' novel features, such as being flexible, stretchable, or self-healing, enable these components to be seamlessly incorporated into the human body and become vital components of BSN.

In [19] a quasi-end-fire spoof surface plasmon polariton (SSPP) leaky-wave antenna (LWA) is presented for wearable application. To account for the deformation that occurs in real-world on-body situations, the LWA is laminated by nonwoven fabric and polyurethane (PU) web on top of a highly flexible textile substrate.

The textile-based LWA generally performs better than traditional designs, which can be attributed to its smaller footprint, robustness against deformations, lower profile, and high-quality radiation pattern. In addition, testing for the design near the human body revealed a SAR of less than 1.6W/kg and a stable impedance-matching performance. The prototype of end-fire antenna on the body for different body bending is shown in Fig. 9.4a whereas their S-parameters are shown in Fig. 9.4b. The spoof surface plasmon polariton transmission line has made it possible to integrate the transmission line on the body textile which helps in reducing the transmission loss. In [20], the Bluetooth sensor module is powered by a wearable radio frequency (RF) energy-localized harvester that has several antennas and a spoof surface plasmon (SSP) structure for multi-beam radiation. Four independent antennas are connected by SSP waveguides in the harvester, and the structure is excited by the loop to produce multi-beam radiation and enhance energy harvesting. The advantage of this topology is that it is free from connectors so no soldering is required which makes it a perfect candidate for wearable applications.

9.4 Choice of Substrate for Flexible or Textile Design

The choice of substrate plays an important role in the design and application of a particular circuit. In recent years as the applications of wireless body area networks such as biomedical implants, and smart watches for health care are increasing, there

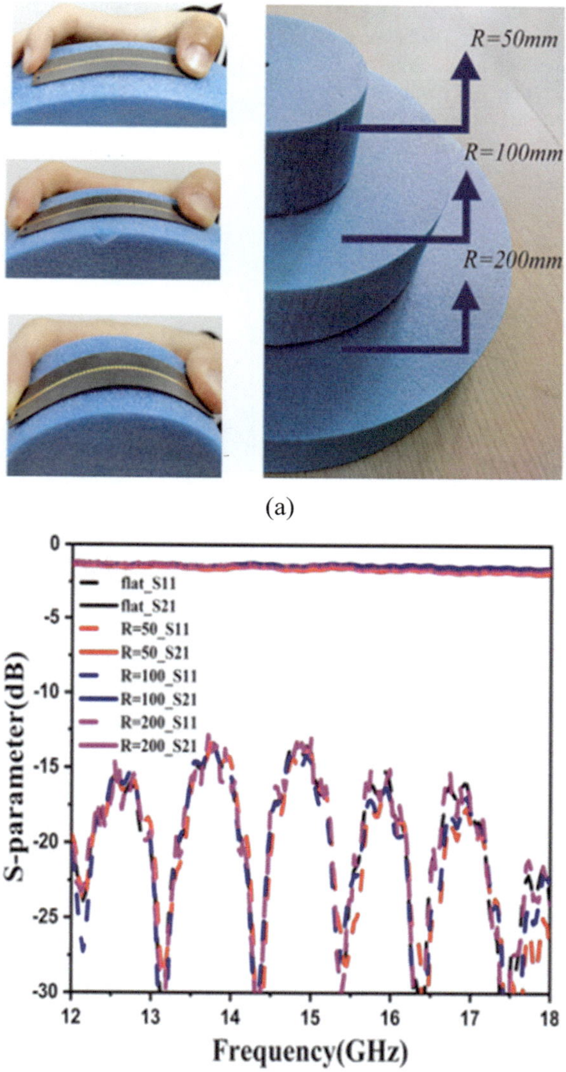

Fig. 9.3 **a** The SSPP transmission line with different bending radius $R =$ 200, 100, and 50 mm. **b** S-parameters of transmission line with different bending radius [18]

has been rapid growth in the advancement of textile materials and flexible electronics components such as electronics textiles or smart textiles. In [21] different textiles and flexible materials are explained with their opportunities and challenges in wireless power transfer systems. The substrates used to design flexible circuits are made of polymers like polyimide and polyethylene terephthalate (PET), cellulose paper composite, polydimethylsiloxane (PDMS), poly vinyl phenol, etc. Polydimethylsiloxane (PDMS) elastomer is widely used in microfluidic sensing with many advantages, flexibility, biocompatibility, durability, etc.

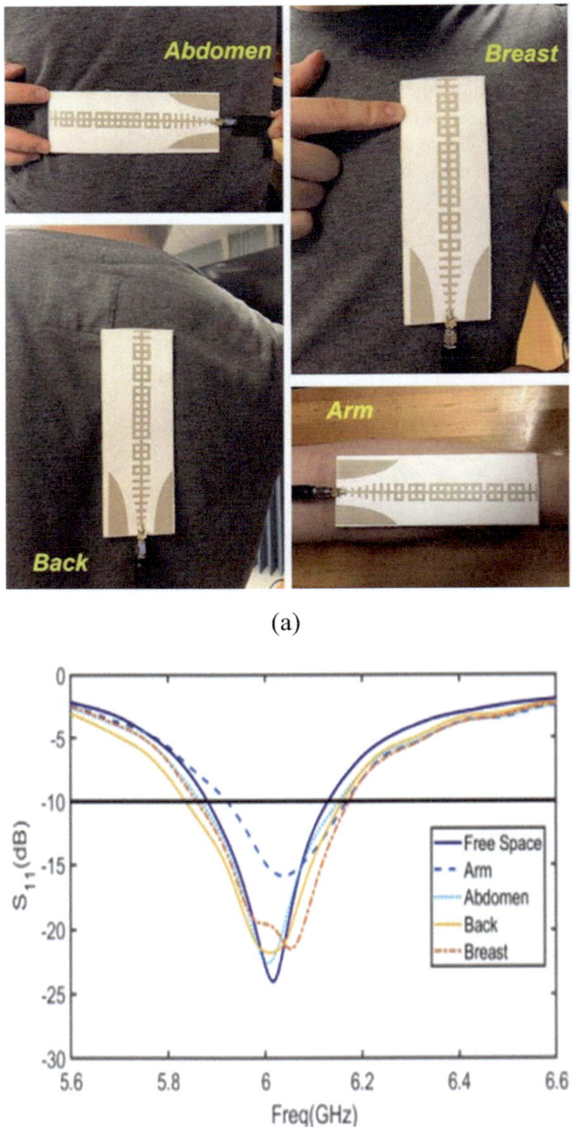

(b)

Fig. 9.4 **a** The end-fire antenna measurement on the body for different body bending. **b** S-parameters for antenna placed on different body bending [19]

A novel approach based on substrate-integrated plasmonic waveguide (SIPW) technology to realize stretchable and flexible microwave (FSM) filters is proposed [22]. A lamp-strip-shaped spoof surface plasmon polariton (SSPP) unit by using a shape-preserving technique and combining them with serpentine interconnections via the 'island-bridge' mechanism to maintain the radio frequency (RF) performance under deformation is created. To achieve the bandpass response, the serpentine patterned substrate-integrated waveguide (SIW) assembled with elastic material has the SSPP structure etched on its upper surface. The PDMS material is used to design waveguide which serves as a biocompatible, mechanically stretchable, and self-healing silicone elastomer. Hence, to provide support and alleviate strain concentration for the conductor, we initiated the preparation of PDMS as the elastic substrate. The prototype covers the X-band and maintains stable performance at conditions of the normal state, 180° bending, 90° twisting, and 120% stretching, and has potential applications in stretchable electronic systems.

Since Kapton polyimide film has a low loss factor across a broad frequency range and a good balance of electrical, chemical, and physical properties, it is used as the antenna substrate. Its resilience, flexibility, low dielectric loss, and thermal endurance are well known. In [23] octagonal and square-shaped coplanar waveguide (CPW)-fed nano silver inkjet-printed Kapton polyimide-based flexible antenna sensor for liquid acetone/water detection is defined. The resonant frequency and the reflection coefficient's amplitude that corresponded to it were monitored for sensing purposes. The fabrication steps clearly define how with the help of an inkjet printer we can print our circuit on a flexible substrate such as Kapton in this case. The suggested sensor can wrap around the sample holder surface and sense near a liquid mixture (acetone and water) because it can function in deformed conditions. Hence, the suggested antenna sensor holds significant promise for noncontact liquid acetone sensing applications, opening up new avenues for antenna-based liquid sensing technologies.

In [24] novel wearable antenna sensor for non-invasive sweat monitoring and communication that is based on localized surface plasmons (LSPs) is explained. To achieve multiband and miniaturization, the sensor is composed of corrugated plasmonic structures, which are based on the traditional printed monopole antenna. The antenna sensor can achieve flexibility and conformality by using polyimide as the dielectric substrate, which keeps it in stable operating states even when bent and deformed. Furthermore, the suggested sensor is extremely sensitive to sweat volume, as evidenced by the substantial shift in the antenna sensor's resonant frequency with sweat volume and concentration.

There are a lot of promising applications for flexible and conformal interconnects in body-worn or wearable electronics as a possible means of transmitting signals. Due to dearth of techniques to simultaneously control wave propagation and achieve the deformable structure, high-efficiency wave propagation and conformal structure deformation around the human body in radio communication continue to face significant challenges. In [25] a new paradigm for building spoof plasmonic interconnects (SPIs) with strong robustness, multifunction performance, and high elasticity is proposed. This paradigm is inspired by the kirigami technology and supports radiofrequency (RF) surface plasmonic transmission.

As shown in Fig. 9.5, stretching from a simply connected metallic surface, the Type-I SPI opens its high-efficiency transmission band by taking advantage of the strong field confinement characteristic of spoof surface plasmons polaritons. In the meantime, the kirigami-based SPI's broadband transmission shows

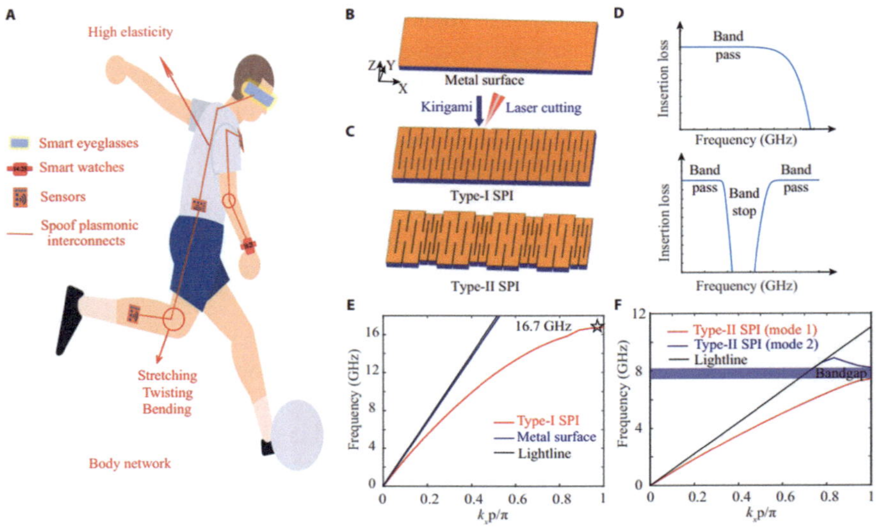

Fig. 9.5 Wireless body NETs based on high-elasticity SPI. **a** Conceptual illustration of the body network based on SPIs. The schematic illustration **b** before and **c** after laser cutting metal surface. **d** Illustrated insertion loss vs. frequency for these 2 stretchable SPIs in (**c**). **e** Dispersion curves of stretchable Type-I SPI and metal surface. **f** Mode 1 (fundamental mode) and mode 2 (second mode) for the stretchable Type-II SPI [25]

remarkable stability and robustness when subjected to intricate deformations like twisting, bending, and stretching. Furthermore, by stretching the buckled structure, the prepared Type-II SPI, which consists of two distinct subunit cells, can achieve band-stop transmission characteristics. Its center frequency can be dynamically adjusted. In the future, wearable RF electronics can function more elastically thanks to these kirigami-based interconnects, which also offer high compatibility for large body motion in body network systems.

Fiber-based devices have gained popularity in textile-based wearable electronics due to their flexibility and unique structure, in response to the growing demand for wearable devices designed for seamless integration with 3D bio-surfaces [26]. Specifically, there has been little reporting on diodes with a fiber structure, even though diodes with rectifying capabilities are essential for a variety of electronic systems. The diode is made up of a LiCl hydrogel electrolyte, a Ti-based fiber cathode on Au nanoparticle-based flexible fiber electrodes, and a double-helical Zn-based fiber anode. Because of the exceptional elasticity and stability of the fiber electrodes, the fiber-based ionic diode maintains its rectifying ability even after being washed and bent significantly.

The materials used in current microwave filters need to be fabricated using complex and costly techniques. Furthermore, they don't work with wearable and flexible platforms. A tunable fabric microwave filter (TFMF) using organic electrochemical transistors (OECTs), which are lightweight and have excellent mechanical

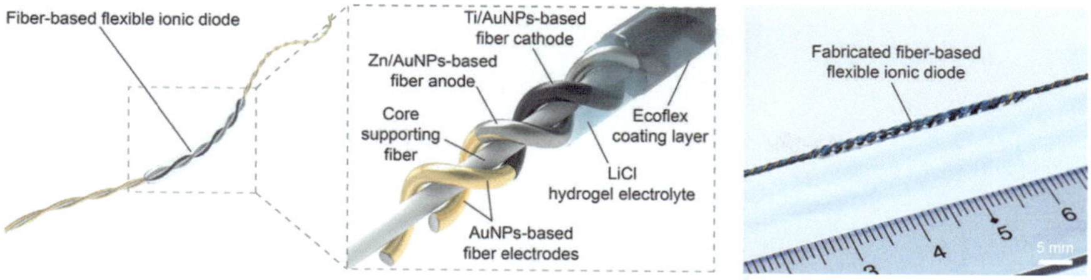

Fig. 9.6 Photograph of the fabricated fiber-based flexible ionic diode [26]

flexibility and deformability is reported in [27]. The approach takes advantage of the spoof surface plasmon polariton (SSPP) structure. The OECT is printed to act as a state-changing material, while the TFMF is made of fabric materials. Superior flexibility, high planar integration, and enhanced wearing comfort are all displayed by the developed TFMF. Furthermore, by using various voltage sequences, the operating frequency of the TFMF with well-designed gradient structures and multi-state dispersion characteristics can be effectively tuned (Fig. 9.6).

The material used as a substrate to design our circuit on textile can be cotton, polyester cotton, denim, woven polyester, felt, polycotton, or organza whereas, for printing structure on them, we use copper tape, conductive fabric, conductive yan, screen printing. These textile circuits find a wide application in RFID and Wearable electronics for different applications [28–31]. In [28], the textile antenna and a small UHF radio frequency identification (RFID) reader are intended to be combined into a smart glove. Smart glove antenna is constructed from conductive threads or by plating metal or an alloy onto non-conductive fabric. In [30], thin camouflaged-military print, cotton jean cloth using conductive copper threads, and sewing embroidery technique are used to create the radiating structure shown in Fig. 9.7. Different measurement positions of the antenna are shown in Fig. 9.8 demonstrating the practical application of the on-body antenna.

Radiation loss under deformations such bending, folding, and twisting is a major problem for conventional flexible circuits. Due to their superior field confinement properties, SSPP TLs may have lot of potential in flexible circuits under deformation. In light of this, a flexible SSPP TL is suggested in this study, and experiments show that it has excellent transmission performance under a range of deformations at operating frequencies between 10 and 20 GHz An experimental demonstration and presentation of a novel kind of flexible TL based on SSPPs are provided. Despite being bent or twisted, the SSPP TL can continue to have low reflection and strong transmission in a broad frequency range. Due to simplicity and inexpensive cost, this SSPP TL offers several applications in future flexible circuits [32].

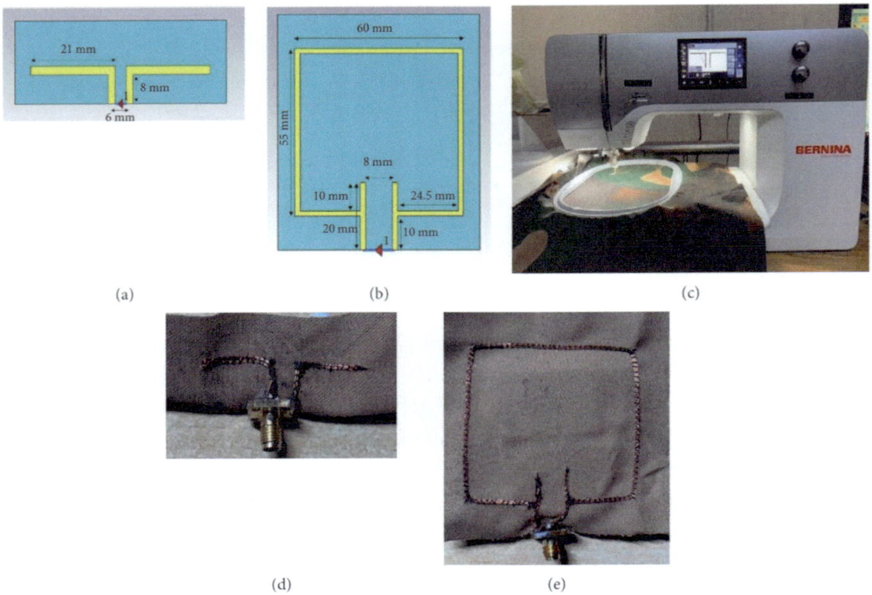

Fig. 9.7 Schematic of the textile-based **a** dipole and **b** loop antenna. **c** Bernina 720 sewing machine used for sewing the radiating element structures in the form of embroidered patterns over the denim-cotton fabric. Fabricated prototype of the textile **d** dipole and **e** loop antenna [30]

The suggested SPP waveguide has horizontal dimensions that may be adjusted between 6.36 and 7.12 mm. Furthermore, in the 150–190 GHz range, the SPP waveguide can produce transmission continuous phase shifts of −30°, −60°, −90°, and −120° with stretchable factors of 1.05, 1.1, 1.15, and 1.2, respectively. Stretchable deformation has no effect on a split ring resonator (SRR), which operates as the equivalent magnetic dipole according to the characteristic mode theory. At 193 GHz, the suggested SPP waveguide produces a tunable passband with a fixed notched frequency when loaded with four SRR cells. In 1–1.2 stretchable states, another stretchable SPP resonator functions as the analogous electric dipole, functioning between 284 and 256 GHz [33].

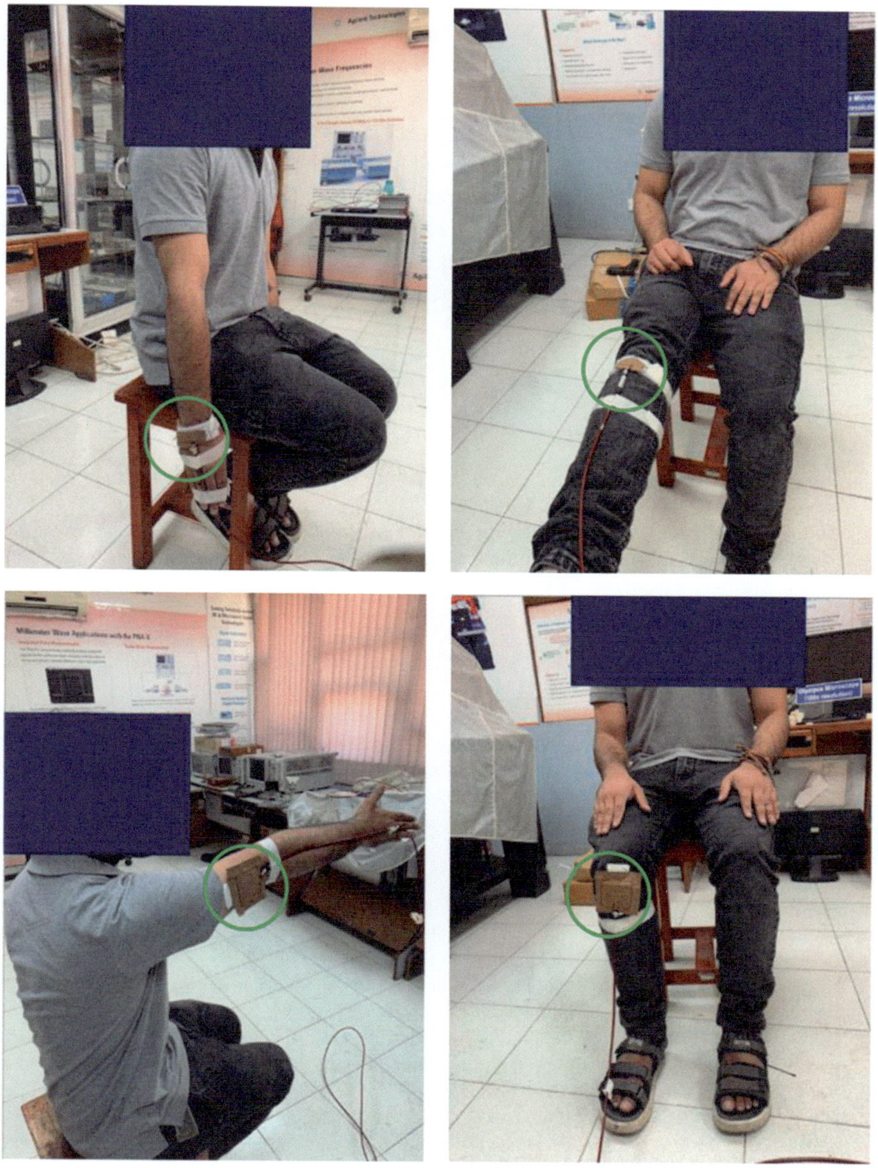

Fig. 9.8 Measurements positions of various on-body antenna locations and bending curvatures for the upper and lower limbs [30]

9.5 Conclusion

In this chapter, we have studied different types of flexible and textile-based circuits for wireless body area networks. The chapter clearly states the need for textiles or flexible antennas in our day-to-day lives and how they are making our lives simple. Different types of flexible or textile substrates are discussed and the methods used to imprint the circuits on flexible substrates are presented. The use of SSPP in the field of flexible or textile substrates is advantageous and helps to resolve the problem of bending loss and cross talk.

References

1. Chang, X.L., Chee, P.S., Lim, E.H.: Compact conformal tattoo-polymer antenna for on-body wireless power transfer. Sci. Rep. **13**, 9678 (2023)
2. Koul, S.K., Bharadwaj, R.: Emerging technologies and future aspects. In: Koul, S.K., Bharadwaj, R., (eds.) Wearable Antennas and Body Centric Communication: Present and Future. Springer, Singapore (2021)
3. Khaleel, H.: Innovation in Wearable and Flexible Antennas. WIT Press, Billerica, MA, USA (2014)
4. Kirtania, S.G., Elger, A.W., Hasan, M.R., Wisniewska, A., Sekhar, K., Karacolak, T., Sekhar, P.K.: Flexible antennas: a review. Micromachines **11**, 847 (2020)
5. Corchia, L., Monti, G., Tarricone, L.: Wearable antennas: nontextile versus fully textile solutions. IEEE Antennas Propag. Mag. **61**(2), 71–83 (2019)
6. Kiourti, A., Lee, C., Volakis, J.L.: Fabrication of textile antennas and circuits with 0.1 mm precision. IEEE Antennas Wirel. Propag. Lett. **15**, 151–153 (2016)
7. Castano, L., Flatau, A.B.: Smart fabric sensors and e-textile technologies: a review. IOP Publ. Smart Mater. Struct. **23**(5), 1–27 (2014)
8. Tang, W.X., Zhang, H.C., Ma, H.F., Jiang, W.X., Cui, T.J.: Adv. Opt. Mater. **7**, 1800421 (2019)
9. Pan, L., Wu, Y., Wang, W., Wei, Y., Yang, Y.: A flexible high-selectivity single-layer coplanar waveguide bandpass filter using interdigital spoof surface plasmon polaritons of bow-tie cells. IEEE Trans. Plasma Sci. **48**(10), 3582–3588 (2020)
10. Lim, Y., An, S.C., Jeong, H.Y., et al.: Multipole resonance and Vernier effect in compact and flexible plasmonic structures. Sci. Rep. **11**, 22817 (2021)
11. Fu, Q., Liu, L., Luo, G.Q.: A flexible millimeter-wave endfire antenna based on spoof surface plasmon polaritons. IEEE Antennas Wirel. Propag. Lett. **23**(1), 29–33 (2024)
12. Yang, X., Tian, X., Zeng, Q., Li, Z., Nguyen, D.T., Ho, J.S.: Localized surface plasmons on textiles for non-contact vital sign sensing. IEEE Trans. Antennas Propag. **70**(9), 8507–8517 (2022)
13. Tian, X., Zhang, M., Ho, J.S.: Robust and high-efficiency wireless body area networks with spoof surface plasmons on clothing. In: 2019 IEEE MTT-S international microwave symposium (IMS), Boston, MA, USA (2019)
14. Tian, X., Yang, X., Ho, J.S.: Energy-efficient and secure wireless body sensor networks with metamaterial textiles. In: 2019 IEEE biomedical circuits and systems conference (BioCAS), Nara, Japan (2019)
15. Ghaddar, A., et al.: Embroidered spoof surface plasmon polariton transmission line for wireless body sensor network. IET Microw. Antennas Propag. **16**(4), 187–193 (2022)
16. Wang, C.M., Xu, W., Pang, C., et al.: Wearable full-textile spoof surface plasmon polariton transmission line for secured communication in wireless body area network. Plasmonics (2024)

17. Liu, Z., et al.: A body-wide RF energy harvester based on textile surface plasmonic antenna array for wearable wireless power transmission. IEEE Trans. Circuits Syst. II Express Briefs **70**(4), 1470–1474 (2023)
18. Wei, J., Li, W., Niu, L.Y., Gao, S., Cui, T.J., Tang, W.: Bendable transmission line and amplifier of spoof surface plasmon polaritons at microwave frequencies. Opt. Express **31**, 755–764 (2023)
19. Wu, Y., et al.: Design of quasi-endfire spoof surface plasmon polariton leaky-wave textile wearable antennas. IEEE Access **10**, 115338–115350 (2022)
20. Liu, Z., Wu, P., Li, G.: A multibeam and surface plasmonic clothing with RF energy-localized harvester for powering battery-free wireless sensor. IEEE Internet Things J. **9**(15), 13955–13964 (2022)
21. Rahman, S., et al.: Wireless power transfer using electronic textiles: a comparative review. J. Eng. Res. ISSN 2307-1877 (2024)
22. Chen, Z.H., et al.: Flexible and stretchable microwave filter based on substrate integrated plasmonic waveguide. IEEE Trans. Microw. Theory Tech. https://doi.org/10.1109/TMTT.2024.3381156
23. Hossain, K., Dang, Y., Cheffena, M.: Nanosilver inkjet-printed CPW-fed flexible antenna sensor for contactless liquid acetone/water detection. IEEE Sens. J. **24**(7), 9933–9945 (2024). https://doi.org/10.1109/JSEN.2024.3361911
24. Zhang, C., et al.: A wearable localized surface plasmons antenna sensor for communication and sweat sensing. IEEE Sens. J. **23**(11), 11591–11599 (2023)
25. Yao, X., et al.: Kirigami-Triggered spoof plasmonic interconnects for radiofrequency elastronics. Research (2024)
26. Woo, S., Kim, H., Kim, J., Ryu, H., Lee, J.: Fiber-based flexible ionic diode with high robustness and rectifying performance: toward electronic textile circuits. Adv. Electron. Mater. **10**, 2300653 (2024)
27. Chen, H., Geng, M.-Y., Chen, Z.-M., Chen, J.-Y., Zhan, J.-L., Yu, B.-Y., Ju, L., Ding, C., Guan, Y.-S., Liu, Z.-G., Lu, W.-B., Li, Q.: Digitally controlled tunable fabric microwave filter based on organic electrochemical transistors. Adv. Mater. Technol. **8**, 2300428 (2023)
28. Singh, R.K., Michel, A., Nepa, P., Salvatore, A., Terraroli, M., Perego, P.: Compact and wearable yagi-like textile antennas for near-field UHF-RFID readers. IEEE Trans. Antennas Propag. **69**(3), 1324–1333 (2021)
29. Sathishkannan, P., Rajesh, K., Singh, K., Ray, K.P., Singh, M.: A compact and flexible dipole antenna to be integrated into military berets. In: 2022 IEEE Microwaves, Antennas, and Propagation Conference (MAPCON), Bangalore, India (2022)
30. Varma, S., Sharma, S., John, M., Bharadwaj, R., Dhawan, A., Koul, S.K.: Design and performance analysis of compact wearable textile antennas for IoT and body-centric communication applications. Int. J. Antennas Propag. **7698765**, 12 (2021)
31. Sharma, S., Singh, R.K., Basu, A., Koul, S.K., Singh, M.: Koch snowflake fractal based compact spoof surface plasmon polariton transmission line. In: 2024 IEEE INC-USNC-URSI Radio Science Meeting (Joint with AP-S Symposium), Florence, Italy (2024)
32. Wei, J., Li, W., Gao, S., Tang, W.: A flexible transmission line of spoof surface plasmon polaritons at microwave frequencies. In: 2022 IEEE 10th Asia-Pacific Conference on Antennas and Propagation (APCAP), Xiamen, China, pp. 1–2 (2022)
33. Shi, X.G., et al.: Phys. Scr. **99**, 085536 (2024)

Index

S. K. Koul and S. Sharma, *Microwave Integrated Circuits Based on Spoof Surface Plasmon Polariton*, Signals and Communication Technology,
https://doi.org/10.1007/978-981-96-9041-1

MIX
Papier aus verantwortungsvollen Quellen
Paper from responsible sources
FSC® C105338

If you have any concerns about our products,
you can contact us on
ProductSafety@springernature.com

In case Publisher is established outside the EU,
the EU authorized representative is:
Springer Nature Customer Service Center GmbH
Europaplatz 3, 69115 Heidelberg, Germany

Printed by Libri Plureos GmbH
in Hamburg, Germany